# Email Architecture, Design, and Implementations

ISBN-13: 978-1511566681

ISBN-10: 151156668X

**Library of Congress Control Number: 2015906058**

**CreateSpace Independent Publishing Platform, North Charleston, SC**

**Kevin Thomas**

# Table of Contents

# Disclaimer

The Author has strived in every way to be as accurate and complete as possible in the creation of this book, notwithstanding the fact that he does not warrant or represent at any time that the contents within the book are accurate due to the rapidly changing nature of the subject.

While all attempts have been made to verify each and every piece of information provided in this publication, the Author assumes no responsibility for any errors, omissions, or contrary interpretation of the subject matter herein.

# Who is This Book Intended For

This book aims at helping system and network administrators and engineers gain a better understanding of E-mail Architecture, its design and implementation.

This book will cover numerous aspects that play a role in the design and implementation of email architectures, such as the ever-so-important protocols, and characteristics of large scale architectures, such as failover, resilience and multiple geographical environments.

The book then moves on to discuss the common types of mail servers along with their description and setup instructions. Popular email clients, DNS setup for email, attachments, and common email services will also be discussed.

As we proceed through the latter chapters of the book, the misuse of email – spam – will be discussed along with ways to avoid it and how organizations can prevent their email from being blacklisted.

Email security is of a paramount importance as there is an abundance of viruses and malware that can compromise data security. The methods of encryption and digital signing are some ways to secure email, and will be looked into in detail in the ninth chapter of this book.

Last, but not the least, etiquettes of email will be talked about in the last chapter of this book.

It is hoped that this book will give valuable insight to all those who seek it.

# Chapter 1: Overview

Electronic Mail, more commonly known as E-mail has transformed considerably in both its scale and complexity over the past three decades. The changes involved have been more evolutionary than revolutionary, thereby indicating a strong desire to conserve its usefulness and the installed base. Today, there is great distinction between hordes of different e-mail operators, the numerous components used for providing services, along with the various types of components used to transfer the messages.

The number of participants in the technical community has increased dramatically, thanks to the greater public collaboration on the technical, operations, and policy aspects of email. In order to be able to collaborate effectively and productively on such a large and complex scale, it is crucial that all participants work from a common perspective and utilize a common language when it comes to describing the components involved in the construction of the email architecture.

## History of Email

As the 1960s progressed, companies started using mainframe and mini computers extensively. This usage brought about the need for users to send messages to each other through terminals. However, it wasn't until 1993 when the email that we now are well aware of came about. The events that occurred between this window are of great importance. Let's go through them to see the impact they made on the modern day electronic mail, and the role they played in the history of email.

## The Advent of Timesharing Computers

As stated, throughout the 1960s, we saw an increase in companies using timesharing computers, due to which there were several instances where research organizations had to use programs for exchanging text messages and even real-time chat. However, this was a fairly limited form of communication as only the group of people using a computer could exchange these messages.

## READMAIL and SNDMSG

The early 1970s saw Ray Tomlinson introduce local email programs termed READMAIL and SNDMSG. This was the time when we saw the "commercial at" symbol being brought to light, as Tomlinson chose to combine the user and host names. It resulted in the modern "user@host" addressing standard which is now used throughout the electronic mail world.

## MLFL and MAIL

The year was 1972 when MLFL and MAIL were introduced. FTP provided a separate copy of the emails being sent to recipients. While also offering other features, the main standout was SMTP protocol which enabled users to send out a single mail to a domain having multiple addressees.

## RD

Steve Lukasik, the then director of the IPTO improved on READMAIL by adding a collection of macros to the Tenex text editor and named this program the RD. The main feature of this emailing tool was the ability to sort email headers by date and subject so users could order their inboxes while also read, delete, and save messages.

## MSG

MSG is arguably the first program that gave us a glimpse of the modern electronic mail. This was a powerful upgrade put forward by John Vittal, which included the options for message forwarding and automated Answer commands. According to modern day experts, the Answer command was a revolutionary introduction as it resulted in an exponential outburst of the use of electronic mail within a short period of time. This was the time when email changed from simply sending independent messages to actually having proper conversations.

## MH/MS

1975 saw Steve Walker, a program manager from DARPA, initiate a project for developing an email capability for the Unix OS which was supposed to resemble MSG. Dave Farber undertook the project, who was a professor at the University of California at Irvine. The system that resulted was supportive of multiple user interfaces, ranging from the basic email command for Unix to the interface of MSG. It came to be known as MS. Despite its prowess, this program was quite slow. Due to this, it saw a follow-up project which resulted in the development of another application called MH (abbreviation of Mail Handler). Ever since 1982, this has been the standard email program for the Unix operating system.

## RFC 822 and RFC 733

In 1977, multiple email program designers collaborated on an initiative by DARPA for collecting a variety of email data formats into one coherent specification. This resulted in RFC 733, which was a combination of existing documentation supported by slight innovations. RFC 822, an upgrade put forward by one of the designers that developed the RFC 733, was introduced in 1982. This was the first instance where the standard for describing the syntax of domain names was used.

## MMDF

Dave Farber and Dave Crocker, the geniuses behind several email-related programs previously, worked together for a project for the U.S. AMC (Army Material Command). Together, they were supposed to develop a program that could relay email over dialup telephone lines for sites that were unable to connect directly to the ARPANET. As a result, the first version of what later came to be known as the Multi-Purpose Memo Distribution Facility (MMDF) was developed. After this, MMDF went through several upgrades and overhauls, but Dave Crocker wasn't the one who brought them about as he left the project, being followed by other developers including the likes of Doug Kingston, Ed

Szurkowski, and Craig Partridge. Some of the new features of the updated versions of MMDF included ISO/CCITT OSI X.400 email standard support, development of a robust TCP/IP layer, and email relay capability for the CSNET.

## Sendmail

Throughout the early 1980s, email relaying was further being carried out through the use of simple UUCP technology. This was being done at the University of California at Berkley, a place where the BSD operating system also found its feet and was developed. An email program called delivermail was created by Eric Allman later on, for the purpose of cobbling together several electronic mail transport services. This was done to create, in effect, an integrated email capability of storing and forwarding. Allman later on used the experience from this creation to establish the sendmail email program. This program was distributed alongside the BSD Unix, and has ever since gone on to establish itself as the most commonly used server for SMTP on the World Wide Web.

## Commercial Mail

The connection between MCI Mail and NSFNET through the CNRI (Corporation for the National Research Initiative) was arranged by Vinton Cerf. This event took place in the year 1988, and the sole reason behind this development was "experimental use". Remarkably, this provided the very first sanctioned commercial usage of electronic mail on the Internet. Soon after, more developments were made in the coming year.

## Online Services

During 1993, the extensive network service providers Delphi and America Online began connecting their proprietary systems for electronic mail to the World Wide Web. This began the adoption of Internet email on a large scale, and soon after it became a global standard.

# A General Overview of the Email Architecture

The infrastructure of an e-mail server comprises of a number of components that work in collaboration to send, receive, relay, store, and deliver e-mail, thus forming the complete e-mail architecture. An e-mail server workload utilizes the standard Internet protocols for the purposes of transmitting and retrieving mail given below:

▸ Simple Mail Transfer Protocol (SMTP)
▸ Post Office Protocol (POP)
▸ Internet Message Access Protocol (IMAP)

The table given below highlights each component of a typical mail server along with its description and a few examples:

| Component | Description | Example |
|---|---|---|
| Mail User Agent (MUA) | An application that allows users to view, create, send, and receive mail. MUA is situated at a client system, such as a PC or workstation. | ▸ Eudora<br>▸ Microsoft Outlook<br>▸ Mozilla Thunderbird |
| Mail Transfer Agent (MTA) | MTA is an application which is used to send, receive, and store mail. This program establishes where and how the email will be stored. | ▸ Microsoft Exchange<br>▸ Postfix<br>▸ Sendmail |

| Mail Delivery Agent (MDA) | MDA in an application which saves the email received to MSA. This application may also be capable of performing extra tasks including filtering of email or diverting email to subfolders. | The Postfix and Cyrus applications have some or all of the functions of an MDA. |
|---|---|---|
| Mail Storage Area (MSA) | MSA is a local system/server email is stored by the MTA. This is the location from where the MSS grabs email as a result of a request from MUA. | ▸ /var/mail/spool/*username*/<br>▸ Maildir<br>▸ Mbox |
| Mail Storage Server (MSS) | MSS is an application which retrieves mail from MSA and then returns it to MUA. | ▸ Cyrus<br>▸ Dovecot |

The figure below shows the components of a mail server along with how the email flows through them:

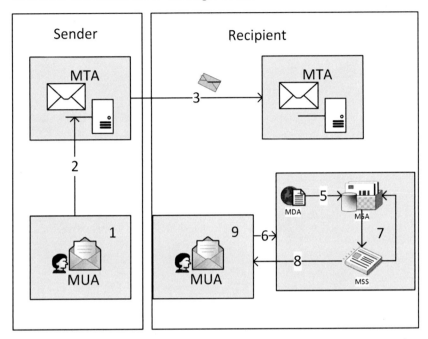

The email passes through the various components of a mail server in the following manner:

1. The sender, from their MUA, creates the email and sends it.
2. The MUA utilizes SMTP to transmit the email to the MTA.
3. The MTA then relays & routes the email correctly to the MTA in the recipient's domain.
4. The recipient's MTA transmits the email to an MDA in the recipient's system.
5. The MDA saves the email in the MSA.
6. The recipient's MUA queries an MSS.
7. The MSS uses POP or IMAP to retrieve email from the MSA.
8. The MSS then returns the email to MUA.

9.   The recipient, from their MUA, views the email sent to them.

Additional applications, such as antivirus, filtering, and anti-spam applications, can be used along with the email clients and the mail server applications to pre-process or post-process the email.

# Chapter 2: Protocols

The interactions between email clients and servers are managed through email protocols. The most common email protocols consist of SMTP, IMAP, POP, and MAPI. The majority of email software available supports one of the three protocols, while some support more than one. It is essential to understand which email protocols exist so that the correct ones can be selected and configured for an email account to function correctly.

## SMTP

Simple Mail Transfer Protocol or SMTP is a type of application layer Internet standard for transmitting email. It was first defined in 1982 by RFC 821, and was updated in 2008 with Extended SMTP additions by the RFC 5321. The last updated version is one that is in widespread use till date.

By default, SMTP utilizes TCP port 25. The protocol used for submission of mail is the same; however, it uses the port 587.

When the SMTP connections are secured using SSL, they are known as SMTPS and use the port 465 by default. Even though mail servers and other mail transfer agents, when sending and receiving mail, use SMTP, client mail applications (at the user-level), use SMTP only for the purpose of sending messages. For receiving mail, these user-level client mail applications use IMAP or POP, as will be discussed below.

Some proprietary systems such as the Microsoft Exchange and numerous web-based email services such as Gmail, Windows Live, and Yahoo mail use their own set of non-standard protocols to access mail accounts on their servers, all of these services use SMTP when sending or receiving email from outside their systems.

## How SMTP Works

The way e-mail is sent and received turns out to be a mystery for many people. In this section, we will look at how the protocol that is responsible for sending e-mail works. As previously mentioned, SMTP listens on port 25; however, it would be more appropriate to say that the SMTP server (rather than the protocol) would be listening on this port for any client connections.

SMTP is, by its design, an application layer protocol. It utilizes Transmission Control Protocol (TCP) as the protocol for transport while using Internet Protocol (IP) for the purpose of routing. Similar to Hypertext Transfer Protocol (HTTP), SMTP also features certain status codes that improve its functionality.

These status codes are employed to relay particular conditions between a client and a server. SMTP does conform to the popular client/server model. For example, Microsoft Outlook is a client and Microsoft Exchange is the server.

In addition to the status codes used by SMTP, there are also a number of SMTP commands. For instance, it uses the 'AUTH' command for authentication and 'EHLO' command for extended hello. It is through these commands that the email client and server are able to communicate. In order to fully understand how this process works, it is necessary to follow the example given below.

The 'HELO' SMTP command is present in the following packet:

```
06/09/2005 06:10:46.595221 192.168.1.100.40565 > 192.168.1.200.25: P [tcp sum
ok] 159505509:159505543(34) ack 578397676 win 33304 <nop,nop,timestamp
310237481 108030715> (DF) (ttl 52, id 34293, len 86)
0x0000   4500 0056 85f5 4000 3406 5235 c0a8 0164      E..V..@.4.R5B...
0x0010   c0a8 01c8 9e75 0019 0981 dc65 2279 a5ec      .....u.....e"y..
0x0020   8018 8218 0449 0000 0101 080a 127d d929      .....I.......}.)
0x0030   0670 6afb 4845 4c4f 2077 6562 3334 3231      .pj.HELO.web3421
0x0040   332e 6d61 696c 2e6d 7564 2e79 6168 6f6f      3.mail.mud.yahoo
0x0050   2e63 6f6d 0d0a                                .com..
```

The 'HELO' command is used after three TCP/IP handshakes have been completed between a client and the server. The

HELO command does exactly what it sounds like, the client says hello to the server. After the HELO command, the domain is mentioned from where the client is present. The domain is in bold for easy identification in the above excerpt.

Once the HELO command has been passed onto the server, the mail client informs the server that it has an e-mail from someone. This can be seen in the excerpt below (MAIL FROM):

```
06/09/2005 06:10:46.641311 192.168.1.100.40565 > 192.168.1.200.25: P [tcp sum
ok] 159505543:159505580(37) ack 578397699 win 33304 <nop,nop,timestamp
310237486 108030720> (DF) (ttl 52, id 35311, len 89)
0x0000   4500 0059 89ef 4000 3406 4e38 c0a8 0164     E..Y..@.4.N8B...
0x0010   c0a8 01c8 9e75 0019 0981 dc87 2279 a603     .....u......"y..
0x0020   8018 8218 053c 0000 0101 080a 127d d92e     .....<.......}..
0x0030   0670 6b00 4d41 494c 2046 524f 4d3a xxxx     .pk.MAIL.FROM:<x
0x0040   xxxx xxxx xxxx xxxx xxxx xxxx 4079 6168     xxxxxxxxxxxx@yah
0x0050   6f6f 2e63 6f6d 3e0d 0a                       oo.com>..
```

SOURCE: http://www.vanemery.com/

Moving back a bit, the TCP sequence numbers can be seen to follow each other in both the packets, just as they should be following. We can deduce that the mail server has not yet acknowledged either of these packets as 'seen'.

It is important to inspect the above packet in detail. So far, we can see that an IP header is present right at the top – which is utilizing IPv4 (Bold and Underlined **4** in the 0x0000 line).

It is also evident that TCP is the transport protocol (Bold and Underlined **06** in IP header in 0x0000 line).

We can see that 12 bytes of TCP options have been set in the TCP header (Bold and Underlined **8** in 0x0020 line).

Starting from 4d41 and proceeding onwards (underlined), this is the actual SMTP application layer data.

It can also be seen that there are two different TCP options available. There is the NOP (no operation) and the timestamp as depicted in the ASCII excerpt above. Let us now analyze these down at the HEX level.

Beginning at the bytes 0101 above, this portion signifies the TCP option 01 and its length of the said option is 01 byte, that is, one byte. After this, there is byte 08, which characterizes the timestamp option. Byte 0a follows this and represents the length of the above timestamp option, in bytes. 0a equals to ten in decimal.

Last but not the least, bytes 127d d92e represents 310237486 as the first timestamp value, followed by bytes 0670 6b00, which signifies the value of the final timestamp value as 108030720.

The timestamp option cannot be seen all the time; this is why it was important to highlight it here.

Getting back to the topic of how SMTP works, after these two packets, what follows is yet another packet containing the actual message. This packet also contains the e-mail body as well as the header fields. The 'e-mail body' refers to the contents of the e-mail (the actual message). Once this has been done, the client severs its connection with the SMTP server by using the QUIT Command.

# SMTP Commands

The following table lists few of the SMTP commands:

| SMTP command | Function of the Command |
| --- | --- |
| **HELO** | Client sends to identify itself. Typically sent with the domain name. |
| **EHLO** | This command allows the server to identify support for Extended Simple Mail Transfer Protocol (ESMTP) commands. |
| **MAIL FROM** | Defines the message's sender; utilized in form MAIL FROM:. |
| **RCPT TO** | Defines the message's recipients; utilized in form RCPT TO:. |
| **TURN** | Enables the client and server to switch their roles and send e-mail in reverse direction without needing to establish a new connection. |
| **ATRN** | The ATRN (Authenticated TURN) command optionally uses one or more of the domains as parameters. The command will have to be rejected if the session is not authenticated. |
| **SIZE** | Provides a mechanism through which the server can specify the maximum supported message size. Compliant servers have to provide size extensions to point out the maximum acceptable size limit. Clients have to make sure they do not send messages larger than the specified size. |
| **ETRN** | This is an extension of SMTP. SMTP server sends ETRN to request another server to send e-mail that it has. |
| **PIPELINING** | Gives the ability to send, without waiting for a response, a stream of commands after a command. |
| **CHUNKING** | An ESMTP command that is used to replace the DATA command. This eliminates the SMTP host's need to scan for the end of the data. A BDAT command is sent by the CHUNKING command along with an argument, which contains the total bytes in the message. The server at the receiving end counts these bytes in a message and, if the message size matches the value that was sent by BDAT command, the server presumes that it has all of the message data. |
| **DATA** | The client sends this command to initiate transfer of the message content. |
| **DSN** | An ESMTP command used to enable delivery status notifications. |
| **RSET** | Used to nullify the message transaction and reset the buffer. |
| **VRFY** | Used to verify that the mailbox is accessible for delivery of message; for instance, vrfy betty is used to verify whether a mailbox for Betty exist on |

| | |
|---|---|
| | local server. |
| **HELP** | Returns with a list of commands supported by the SMTP service. |
| **QUIT** | Used to terminate a session. |

SOURCE: http://www.vanemery.com/

# POP

Post Office Protocol or POP is an Internet standard protocol utilized by email clients when retrieving email from a server over the TCP/IP connection. POP has undergone development through several different versions, with the latest one being the version 3.

Almost all email clients and services in operation today support the POP3.

POP provides simple download and delete method for accessing mailboxes located on remote servers. While there is an option to leave a copy of the mail on the server, email clients using POP generally download all the email from a server and save the email locally on the computer. A POP3 server is designed to listen on port 110. After protocol initiation, encrypted communications are requested using the STLS command, or in some cases, by POP3S, which establishes a connection with the server using SSL or TLS, on TCP port 995.

When the POP session opens up the mailbox, the messages available to the client are fixed, and are identified by a message number unique to the session. Alternatively, each of the messages can also be assigned a unique identifier by the POP server. This permanent, unique identifier allows the same message to be accessed in different POP sessions. After the mail is retrieved, the mail is marked for deletion by its message number. Once the client quits the session, the mail that is marked for removal is deleted.

## Protocol Details

POP3 strictly adheres to a client-server model. The client establishes a connection with the server, while issuing text commands; the server responds appropriately. The server itself does not initiate anything at all. As a matter of fact, TELNET can be used to check mail or perform basic troubleshooting by connecting to the POP3 server on the port 110. After the connection has been established, simple commands can be issued using the keyboard and the responses can be viewed.

The only complications in this protocol are due to the additions that were made recently to support new authentication and encryption methods.

The following is a summary of POP3 commands & the responses:

| Minimal POP 3 Command | |
|---|---|
| USER name | Valid in the AUTHORIZATION state |
| PASS string | |
| QUIT | |
| STAT | Valid in the TRANSACTION state |
| LIST [msg] | |
| DELE msg | |
| NOOP | |
| RSET | |
| QUIT | |
| Optional POP 3 Command: | |
| APOP name digest | Valid in the AUTHORIZATION state |
| AUTH [auth type] | |
| TOP msg n | Valid in the TRANSACTION state |
| UIDL [msg] | |

| POP3 Replies: | |
|---|---|
| +OK | |
| -ERR | |
| Extended Commands: | |
| CAPA | Valid in the AUTHORIZATION or TRANSACTION state |
| STLS | Valid in the AUTHORIZATION state |
| Extended Response Codes: | (enclosed in square brackets followed by – ERR reply) |
| LOGIN-DELAY | |
| IN-USE | |
| SYS/TEMP | |
| SYS/PERM | |
| AUTH | |

SOURCE: http://www.vanemery.com/

## Authentication & Security

The one thing that makes POP3 extremely difficult to administer and configure (both as a client and a server) is the multiple number of ways used to authenticate users and encrypt traffic. The clients and the servers have different types of capabilities, and the POP3 protocol has been supplemented over the years ever since its inception, to handle new forms of encryption and authentication. For instance, the CAPA command enables the client to determine what is supported by the server without probing it.

The USER and PASS commands are the most basic methods of authentication. All the POP3 clients out there support them. Due to the fact that these commands pass both the username and the password in an 'open' way, these login credentials can be easily captured when being transmitted. When Linux/Unix accounts with shell access are considered, this can be a major security issue. This is exactly why APOP was developed. APOP is a method that allows sending of password in a secure, hash-based way. The hash continuously changes with each login, thereby rendering reply attacks useless. Nevertheless, there are three drawbacks when APOP is considered, and they are as follows:

- A separate username and password database has to be in existence in plain-text format. This cannot be the same as the Linux/UNIX password file (which is encrypted).

- Some popular clients, such as the Outlook Express, do not have support for APOP

- APOP does not encrypt email

It may be advantageous to separate the username and password database from the host OS's authentication database. For example, of the purpose of a particular server is to handle email only, then it can be easy to administer and manipulate the plain-text database using scripting tools. According to some people, this is a huge security risk because the attacker that gains access to the list will have every user's password. However, if someone already has gained access to the root, this is big enough of trouble by itself. Root has the ability to read mail spools, access an encrypted username/password database, and easily crack them once offline. On the other hand, if a particular email user's POP3 password is the same as the password of their system, then it can be taken from a PC through various methods that are used to logon to corporate servers.

With some types of the SASL-based AUTH methods, the authentication database is separate from that of the host OS. In some scenarios, it would be ideal to use APOP or AUTH CRAM-MD5; however, in others, it may be more flexible to use plaintext passwords with PAM.

Outlook Express has no support SASL CRAM-MD5 or APOP, this is why many Internet Service Providers (ISPs) were not able to authorize use of such mechanisms. This means that the choice remains between supporting a single method or multiple authentication database method. The least that can be done is to use USER/PASS in conjunction with (or without) TLS/SSL.

While it is clear that it is not a wise idea to send username and passwords openly over a network, mandating TLS/SSL would be a clever strategy. However, this has its own set of disadvantages:

- Depending on how many users and requests per minute there are, organizations may require needlessly powerful systems and hardware SSL acceleration

- Organizations may need to implement support for TLS (port 995) as well as STARTTLS (port 110)

- Organizations will need to purchase certificates and deal with the crypto certificate management

- By using TLS, users might get a false sense of security. TLS is not an alternate to mail encryption tools such as S/MIME or OpenPGP

One question is raised: why should someone run TLS on port 995 as well as 110? This is because many clients have support for STARTTLS (STLS command) on the port 110, whereas other clients have support for TLS on port 995 only. The IETF is trying to push the STARTTLS method for all of the client-server protocols because it eases the ongoing demand for ports. For instance, originally, port 465 was the selected port for TLS support for SMTP. However, the idea was dropped and this port was assigned to some other application. Instead, STARTTLS was made the approved encryption method over SMTP,

Implementing and administering support for both the methods would be a headache for the network administrator, to say the least. The server software may also be needed to be recompiled to ensure that plaintext methods are not made available unless or until TLS has been established.

# Firewall Concerns

POP3 works fine through NAT/PAT devices and firewall devices, it uses port 110 and port 995 for secure POP3.

# POP3 Client Implementations

The following table will list information about some POP3 clients along with their support for certain ports/methods:

| Client | Operating System | APOP | SASL CRAM-MD5 | SASL LOGIN | STARTT LS | TLS 995 |
|---|---|---|---|---|---|---|
| Evolution | Unix/Linux | Yes | Yes | Yes | Yes | Yes |
| KMail | Unix/Linux | Yes | Yes | Yes | Yes | Yes |
| Sylpheed | Unix/Linux | Yes | No | No | Yes | Yes |
| Fetchmail | Unix/Linux | Yes | Yes | No | Yes | Yes |
| Mozilla Mail | Linux/Win/Mac | Yes | Yes | Yes | No | Yes |
| Opera M2 | Linux/Win/Mac | Yes | Yes | Yes | Yes | Yes |
| Eudora | Win/Mac | Yes | No | No | Yes | Yes |
| Outlook Express 6 | Win | No | No | Yes | No | Yes |

SOURCE: http://www.vanemery.com/

# IMAP

Internal Message Access Protocol or IMAP is a protocol designed for email retrieval and storage back in 1986 by Mark Crispin at the Stanford University. It was designed as an alternate to POP and allows establishment of concurrent connections to the same mail account from different machines. Flags can be stored on the server, which allows different clients who are accessing the same account to detect the changes made by others.

IMAP version 4 revision 1 is the latest version of IMAP and is defined by the RFC 3501. An IMAP server normally listens on the famous port 143. IMAP can also be used in conjunction with SSL; hence it is termed IMAPS in such a case and uses port 993.

IMAP allows online and offline operations and email clients that use IMAP to access email typically leave a copy of the message on the server until the user implicitly marks them for deletion.

IMAP offers several advantages over POP, and they are as follows:

- ✓ Supports for online and offline operation modes
- ✓ Allows multiple clients to access the mailbox simultaneously
- ✓ Allows access to partial fetch and MIME message parts
- ✓ Allows access to message state information through flags stored on the server
- ✓ Allows creation/deletion of multiple mailboxes on the server

## The Workings of IMAP

The IMAP protocol follows the basic flow given below to retrieve mail:

1. Open Connection
2. Authenticate
3. SelectFolder
4. Fetch
5. Logout

## 1. <u>Open Connection</u>

To establish a connection between a server and a client, the *Socket* object has to be initialized. For example, the **HigLabo.net** project contains the *SocketClient.cs*:

```
protected Socket GetSocket()

{

    Socket tc = null;
    IPHostEntry hostEntry = null;
    hostEntry = this.GetHostEntry();
    if (hostEntry != null)

    {

        foreach (IPAddress address in hostEntry.AddressList)

        {
          tc = this.TryGetSocket(address);
          if (tc != null) { break; }
        }

    }
    return tc;

}
private Socket TryGetSocket(IPAddress address)
{
        IPEndPoint ipe = new IPEndPoint(address, this._Port);
        Socket tc = null;
```

```
    try
    {
        tc = new Socket(ipe.AddressFamily, SocketType.Stream,
ProtocolType.Tcp);
        tc.Connect(ipe);
        if (tc.Connected == true)
    {

            tc.ReceiveTimeout = this.ReceiveTimeout;
tc.SendBufferSize = this.SendBufferSize;
tc.ReceiveBufferSize = this.ReceiveBufferSize;

        }
}
catch

        {
        tc = null;
        }
        return tc;
        }
        private IPHostEntry GetHostEntry()
        {
        try
        {
        return Dns.GetHostEntry(this.ServerName);
}

    catch { }

    return null;

}
```

After this, you may send a command through the Socket object. The example given below:

The response after establishing a connection is as follows:

**220 mx.google.com ESMTP pp8sm11319893pbb.21**

The format of a single-line response is as follows:

---

[responseCode][whitespace][message]

---

The code 220 indicates ServiceReady.

Now send the hello command:

---

helo example@example.com

---

Command format can be requested as follows:

---

250-mx.google.com at your service, [61.197.223.240]
mx.google.com at your service, [61.197.223.240]
SIZE 35882577
8BITMIME
AUTH LOGIN PLAIN XOAUTH
ENHANCEDSTATUSCODES

---

SOURCE: http://www.codeproject.com/

As shown above, the response message lists the authentication types that are supported by the server. In this case, AUTH, LOGIN, PLAIN, and X0AUTH are supported.

## 2. Authenticate

The next step involves authentication of your mailbox with the correct login details. Here is an extract from a sample *ImapClient.cs* file:

```
public ImapCommandResult ExecuteLogin()

{

    if (this.EnsureOpen() ==
ImapConnectionState.Disconnected) { throw new
MailClientException(); }

    String commandText = String.Format(this.Tag + " LOGIN
{0} {1}", this.UserName, this.Password);
    String s = this.Execute(commandText, false);
    ImapCommandResult rs = new
ImapCommandResult(this.Tag, s);
    if (rs.Status == ImapCommandResultStatus.Ok)
    {
    this._State = ImapConnectionState.Authenticated;
    }
    else
    {
    this._State = ImapConnectionState.Connected;
    }
    return rs;

}
```

SOURCE: http://www.codeproject.com/

### 3. SelectFolder

After the authentication process, a folder has to be selected from where the mail is to be retrieved. In order to select a folder, you would need to get the folder list existing in the mailbox. This folder list can be retrieved by sending a 'list' command to the server.

```
public ListResult ExecuteList(String folderName, Boolean recursive)
{
    this.ValidateState(ImapConnectionState.Authenticated);

    List<ListLineResult> l = new List<ListLineResult>();
    String name = "";
    Boolean noSelect = false;
    Boolean hasChildren = false;
    String rc = "%";
    if (recursive == true)
    {
        rc = "*";
    }
    String s = this.Execute(String.Format(this.Tag + " LIST \"{0}\" \"{1}\"", folderName, rc), false);
    foreach (Match m in RegexList.GetListFolderResult.Matches(s))
    {
        name = NamingConversion.DecodeString(m.Groups["name"].Value);
        foreach (Capture c in m.Groups["opt"].Captures)
        {
            if (c.Value.ToString() == "\\Noselect")
            {
                noSelect = true;
            }
            else if (c.Value.ToString() == "\\HasNoChildren")
            {
                hasChildren = false;
```

```
        }
        else if (c.Value.ToString() == "\\HasChildren")
        {
            hasChildren = true;
        }
    }
    l.Add(new ListLineResult(name, noSelect,
hasChildren));
  }
  return new ListResult(l);
}
```

SOURCE: www.codeproject.com

The command sent to the server is as follows:

---

tag1 LIST "" "*"

---

The server responds in the following manner:

---

```
* LIST (\Mail) "/" "INBOX"
* LIST (\Mail) "/" "Notes"
* LIST (\Noselect \Mail) "/" "[Gmail]"
* LIST (\Mail1) "/" "[Gmail]/All Mail"
......
* LIST (\Mailbox) "/" "[Gmail]/Trash"
tag1 OK Success
```

SOURCE: www.codeproject.com

---

Through the ImapClient class, you can get all the folders through *GetAllFolders* method.

---

```
MailMessage mg = null;

using (ImapClient cl = new ImapClient("imap.gmail.com"))
{
    cl.Port = 993;
    cl.Ssl = true;
    cl.UserName = "xxxxx";
    cl.Password = "yyyyy";
    var bl = cl.Authenticate();
    if (bl == true)
    {
        //Get all folder
        var l = cl.GetAllFolders();
    }
}
```

SOURCE: www.codeproject.com

Before you can receive mail from a server, you have to select a particular folder from within your mailbox. In order to select a folder, use the *select* command.

The following is the *ExecuteSelect* method of **ImapClient** class:

```
public SelectResult ExecuteSelect(String folderName)
{
    this.ValidateState(ImapConnectionState.Authenticated);
    String commandText = String.Format(this.Tag + " Select
{0}", NamingConversion.EncodeString(folderName));
    String s = this.Execute(commandText, false);
    var rs = this.GetSelectResult(folderName, s);
    this.CurrentFolder = new ImapFolder(rs);
    return rs;
}
```

SOURCE: www.codeproject.com

The client sends the follows to the server:

**tag1 Select INBOX**

The server returns with the following response:

**\* FLAGS (\Answered \Flagged \Draft \Deleted \Seen $Forwarded)**
**\* OK [PERMANENTFLAGS (\Answered \Flagged \Draft \Deleted \Seen $Forwarded \*)] Flags permitted.**
**\* OK [UIDVALIDITY 594544687] UIDs valid.**
**\* 223 EXISTS**
**\* 0 RECENT**
**\* OK [UIDNEXT 494] Predicted next UID.**
**tag1 OK [READ-WRITE] INBOX selected. (Success)**

SOURCE: http://www.codeproject.com/

## 4. <u>Fetch (GetMessage)</u>

After selecting the appropriate folder, the mail list can be accessed using the *Fetch* command. The actual mail data can be retrieved through the *GetMessage* method from ImapClient class. A sample code used to receive a message is shown below:

```
private static void ImapMailReceive()
{
  MailMessage mg = null;

  using (ImapClient cl = new
ImapClient("imap.gmail.com"))
  {
    cl.Port = 993;
    cl.Ssl = true;
    cl.UserName = "xxxxx";
    cl.Password = "yyyyy";
    var bl = cl.Authenticate();
    if (bl == true)
    {
      //Select folder
      var folder = cl.SelectFolder("[Gmail]/All Mail");
      //Get all mail from folder
      for (int i = 0; i < folder.MailCount; i++)
      {
        mg = cl.GetMessage(i + 1);
      }
    }
  }
}
```

SOURCE: http://www.codeproject.com/.

The GetMessage method's index must be greater than 1, and not 0.

# MAPI

Messaging Application Programming Interface or MAPI is a messaging architecture along with being a Component Object Model-based API for Microsoft Windows. MAPI enables client programs to successfully become messaging-enabled, -based, or –aware by calling the MAPI subsystem routines which interface with numerous messaging servers. While MAPI has been intended to be independent of a protocol, it is typically used with MAPI/RPC – a proprietary protocol used by Microsoft Outlook to establish communications with MS Exchange.

Simple MAPI consists of a subset comprising of 12 functions. These functions allow developers to add the basic messaging functionality. Extended MAPI provides complete control over a messaging system on client site, including creation & management of messages as well as the management of service providers and client mailbox.

Simple MAPI comes with Windows, as Windows Mail/Outlook Express. On the other hand, Extended MAPI comes with Microsoft Exchange and Office Outlook.

MAPI includes numerous facilities to access message stores, message transports, and directories.

## MAPI Architecture

The basic conceptual components of MAPI comprise of a MAPI server and a MAPI client. The two components work in conjunction with each other in order to create, transport, and store the messages in a MAPI system.

The MAPI clients are designed to deal with three objects:

1.  Addresses

2.  Storage Folders

3.  Attachments and Messages

Each of the objects listed above are implemented slightly differently in numerous MAPI set-ups. For instance, the MAPI OCX tools that come with Visual Basic provide limited access to address and folder information. On the other hand, the Messaging OLE layer (through MAPI 1.0 SDK) provides greater access to both addresses and folders, however, only the full version of MAPI 1.0 empowers programmers with the ability to add, delete, and edit folders & addresses down at the client level.

The MAPI server takes care of the following objects:

1. Address books

2. Message transport

3. Message stores

The client is responsible for creation and manipulation of messages, whereas, the server focuses on transporting the messages. When the client is involved in accessing the storage folders, the server deals with message storage, but both client and server have to deal with the message addresses.

Nevertheless, the MAPI server is responsible for managing the storage, transport, and addressing aspects of messages from numerous client applications. In addition to maintaining a message base for all the available local clients, the MAPI servers also have to move messages from and to remote clients and servers.

## The MAPI Client

The MAPI Client is an application that is designed to run on a user's workstation. This application initiates requests services from MAPI server. These 'client' applications can be generic, such as Microsoft Mail or the Exchange Mail Client. These applications may also be message-aware such as the MS Office applications.

Each of these applications tend to provide access to a server through the 'send menu' option or the command button. Last but not the least, message-enabled applications that use MAPI services can also be built to meet a specific need. Such types of programs present data-entry screens, which ask for message related information, only to format them and send messages through the message server available.

All MAPI clients have to cater for three basic objects as follows:

1. Addresses
2. Storage Folders
3. Attachments and Messages

Depending on the client application's type, a single or multiple number of these objects might be hidden from the client's interface. While they may not be visible to the user, they would still be present as a part of a MAPI architecture.

Let us now describe each of these objects:

## Addresses

Every message has a minimum of two (2) address objects:

- The sender object; **and**

- The recipient object

MAPI offers the ability to add a number of recipient addresses to a particular message. Each of the address object possesses unique properties, and the following table reveals some of the properties of the MAPI Address object.

| Property Name | Type | Description |
|---|---|---|
| **Address** | String | This is the unique electronic address for this address object. The combination of the Typeproperty (see below) and the Address property creates the complete MAPI address. Sample address properties are MikeA@isp.net-Internet address /MailNet1/PostOfc9/MCA-MS Mail address |
| **DisplayType** | Long | The MAPI service allows programmers to define addresses by type. This means you can sort or filter messages using the DisplayType property. Sample address types are mapiUser-Local user mapiDistList-Distribution list mapiForum-Public folder mapiRemoteUser-Remote user |
| **Name** | String | This is the name used in the Address book. Usually, this is an easy-to-remember name such as "Fred Smith" or "Mary in Home Office." |
| **Type** | String | This value contains the name of the message transport type. This allows MAPI to support the use of external message transport services. Sample address types are MS:-Microsoft Mail transport SMTP:-Simple Mail Transport Protocol MSN:-Microsoft Network transport |

SOURCE: http://web.info.uvt.ro/

## **Storage Folders**

MAPI messages may be saved in folders, as many storage folders are defined by the MAPI model. The following storage folders are defined within the MAPI:

- Inbox

- Outbox

- Sent

- Deleted

- User-defined folders

Not all of the MAPI implementations have support for the folders mentioned above. For instance, a Simple MAPI interface provides access only to the Inbox (through .Fetch method of MAPISession Control) and the Outbox (through .Send method of MAPISession Control). Access to any other folder is not allowed.

On the other hand, the OLE Messaging library provides complete access to the entire MAPI storage system. Both the Inbox and Outbox folders can be accessed by name, while the Sent and the Deleted are also available.

MAPI storage folders have been defined as a certain hierarchy. The MAPI model enables creation of a number of levels throughout this hierarchy, including the creation of sub-folders. For instance, the main Inbox folder may have various sub-folders, such as Read, Unread, Urgent, and Other. Additionally, this enables the users to better organize their message stores according to their preference and use.

## Attachments and Messages

The MAPI system moves messages from one location to the other. Thus, at the heart of the MAPI system lies the MAPI message object. All message objects contain two components: message header & message body.

The message header includes information that is utilized by the MAPI to correctly route and track movement of the message being transported.

The message body includes the message portion of a particular message body. Besides these two components, a message object may have a single or multiple attachments. These attachments can consist of any executable program, a binary image, or an ASCII text file.

## The Message Header

All of the information that is needed for the delivery of a message along with its attachments is included in the message header. Depending on the type of messaging service provider, the data stored within the message header differs. Even though the precise items and the values vary between each type of messaging system (MAPI, OLE Messaging or CMC), the basic set of information found in the headers is the same. The table below highlights such information:

| Property Name | Type | Description |
| --- | --- | --- |
| **Recipients** | Recipients object | E-mail address of the person who will receive the message. This could be a single name, a list of names, or a group name. |
| **Sender** | AddressEntry object | E-mail address of the person who sent the message. |
| **Subject** | String | A short text line describing the message. |
| **TimeReceived** | Variant (Date/Time) | The date and time the message was received. |
| **TimeSent** | Variant (Date/Time) | The date and time the message was sent. |

SOURCE: http://web.info.uvt.ro/

## The Message Body

The message body includes the text that is being sent by the sender to the recipient. In most cases, this is simply a text message in pure ASCII format. Some service providers, however, support rich-text format in the message body, thereby allowing use of different colors, fonts, and format codes.

## Message Attachments

All types of Microsoft MAPI support message attachments. A MAPI attachment can be any type of a data file, including binary programs, text, graphics, and others. These are sent with the message body and header. Upon receiving it, the recipient can view, store, and manipulate the attachment on the local computer, depending on the features of the client software.

# The MAPI Server

The MAPI server effectively handles all of the messaging traffic by the MAPI clients. It is usually a dedicated workstation connected to a network; however, this is not a necessity. Microsoft has support for two dedicated MAPI servers:

- Microsoft Exchange Server

- Microsoft Mail Server (eventually replaced by MS Exchange Server)

All MAPI servers are designed to handle the following tasks:

1. Address books
2. Message transport
3. Message stores

## Address Books

The MAPI servers have address books containing all directory information about a specific addressee. The MAPI client can access a number of address books at a given time, due to their independent nature under a MAPI model.

## Message Transport

The process of moving messages from one point to another is known as *message transport.* Based on the MAPI model, the process is separate and distinct, MAPI 1.0 allows utilization of *external message transport.* Putting it in simpler words, this empowers developers and programmers to write software that can handle a certain type of message format and register this transport method within the MAPI system.

## Message Stores

Message stores provide a filing system for all the messages that have been received through message transport. According to the MAPI model, the message store must have a hierarchical format, one that allows for a multilevel storage. There is no limit whatsoever to the number of folder levels allowed to be created in the message store.

# Chapter 3: Common Mail Servers Descriptions and Setup Instructions

When it comes to choosing a mail server for setting up the email architecture, network engineers and administrators are often faced with a number of factors. The selection of a mail server is significantly influenced by the software's technical and functional assessment.

## Selection of Mail Server Products

The factors that are involved during the selection process are as follows:

- Product functionality
- Product architecture
- Capacity scalability/planning
- Platform requirements
- Migration simplicity
- Ease of installation
- Support
- Upgrades/Updates
- Security
- User base

## Product Functionality

Mail server software has evolved considerably over the past few years. The expectation of features has changed considerably. Today, the majority of mail servers available provide more functionality than just sending and accessing messages.

Most of them now even offer the same type of core features; for instance, the support for most common protocols including POP, IMAP, SMTP, as well as Web Mail access. The decision, however, becomes more difficult when particular functional requirements are to be considered. While it is true that almost all the mail servers available do provide support for basic mail access protocols, there are generally slight differences in the functionality offered by each of the mail servers. In such circumstances, it is important to draw up a matrix of your requirements to be able to figure out which product is the most suitable for your needs.

## Capacity Scalability/Planning

Scalability is a major factor, particularly when a mail server is being selected for larger mail architectures. For those organizations that need an email server for a few hundreds of users, scalability is not a major factor. However, if the number of users goes in the number of thousands, than the scalability factor becomes a necessity.

In terms of the capacity planning, the sizing requirements of a certain messaging platform take the following factors under consideration:

- ▸ The number of services required
- ▸ The level of concurrent service usage
- ▸ The performance of IO subsystem
- ▸ The level of logging needed

With that said, it is much better to base your selection on the published hardware specifications & number of users supported by a mail server.

## Platform Requirements

The majority of mail server solutions designed for enterprises often require powerful, expensive hardware and servers. These hardware requirements have to be taken into account when selecting a particular type of mail server.

## Migration Simplicity

It can be quite costly to migrate from one mail server to the other. Depending on the size of the server, it takes at least a day or so to plan and carry out the migration process.

The ease of migration through migration tools is yet another crucial factor that should be looked into.

## Ease of Installation

The mail server should be relatively simple to set up and manage. Many servers come with default installation settings that allow it to be configured in a way that it is ready to be used as a mail server right out of the box – something which is great for network administrators who are relatively new to main server configuration.

## Support

Product support is a crucial factor when it comes to mail server products. This is particularly true in the case of commercial server software as it counts in figuring out the Total Cost of Ownership. Some developers provide support on a per-incident basis, while some offer it on a contractual basis. The cost of support and the response times are also major influencing factors.

## Upgrades/Updates

The mail server software should also receive regular updates to ensure that the users are getting the latest features.

Software that comes with options of upgrading to newer/more powerful versions tends to be a bonus for those organizations that need scalable solutions.

## Security

Security is a critical factor, especially for mail servers. These servers need to be protected from abuse by unauthorized persons and spammers, not to mention the protection against unauthorized entry that could compromise data security.

## Types of Mail Server Products Available

There are mainly two types of mail server products available: **commercial**, and **open-source**.

We will provide an overview of some commercial and open-source mail servers in the upcoming sections to help concerned individuals better understand what to expect from today's modern mail server applications.

## Commercial

### 1. Microsoft Exchange Server (MXS)

Microsoft Exchange Server (MXS) is a mail server, contact manager, and calendaring software designed by Microsoft. It runs on Microsoft Windows Server. The Exchange Server has a history dating back to the early 1990s when email began to evolve into an application highly critical to businesses. This led to the development and periodic releases of MS Exchange Servers over the years, with Microsoft Exchange Server 2003 being the latest version.

The latest version of MXS enables people to deliver email, calendar, or contacts to a mobile device, PC, or a browser.

**Features Offered By Microsoft Exchange Server**

▸ Exchange Server 2013 provides always-on communications while providing full control to the network engineers.
▸ The Exchange building block model is designed to simplify deployment at all kinds of scales, while offering

client load balancing, high availability, and enhanced cross-version interoperability.

▸ Provides greater uptime by offering fast failover times and offers support for many databases-per-volume.

▸ A built-in monitoring and availability solution equips self-healing features that allow automatic recovery from failures.

▸ Easy-to-use load balancing options enhance flexibility & scale at a lower cost.

▸ Allows use of larger, cheaper disks to reduce costs associated with storage.

▸ The built-in Data Loss Prevention capabilities protect sensitive data from being mistakenly forwarded to unauthorized people.

▸ Custom instructions are provided for setting up of server and deployment.

For more information on Microsoft Exchange Server 2013, visit https://products.office.com/en-us/exchange/microsoft-exchange-server-2013.

## 2. Novell GroupWise

GroupWise is a collaboration and messaging platform designed and developed by Novell, which supports email, personal information management, calendaring, document management, and instant messaging. GroupWise platform comprises of desktop client software (available for Windows, Linux, and Mac OS X), and server software (available for Windows Server & Linux).

GroupWise also offers support for WebAccess - a webmail client based on browser. Meanwhile, it enables access to calendaring, messaging, contacts, and other types of data through mobile devices, such as tablets and smartphones. This support is implemented through Novell Data Synchronizer server software via ExchangeActiveSync protocol.

The Novell Messenger handles enterprise-level IM, which can be integrated into GroupWise. GroupWise 2014 is the latest generation of this platform.

The minimum system requirements for installation of Novell GroupWise 2014, as depicted on Novell's website, are as follows:

▶ x86-64 processor (GroupWise runs like a 32-bit application even on a 64-bit processor)

Any one of following operating systems for running GroupWise agents (Post Office Agent, , Document Viewer Agent, Message Transfer Agent , Internet Agent, Monitor Agent):

▶ Novell Open Enterprise Server (OES) 11, with the latest Support Pack
▶ SUSE Linux Enterprise Server (SLES) 11, with the latest Service Pack
▶ Windows Server 2008 R2, Windows Server 2012, or Windows Server 2012 R2, with the latest Service Pack
▶ Sufficient server memory as needed by the OS. Additional memory may be needed depending on the probable load on GroupWise agents.

▸ Sufficient server storage space as needed by each of the GroupWise components.

At least one (or more) of the following directories:

▸ NetIQ eDirectory 8.8.7 or later, plus the latest Support Pack, with LDAP enabled (optional)
▸ Native GroupWise directory (internal; required)
▸ Microsoft Active Directory (optional)

## 3. IceWarp Mail Server

IceWarp Mail Server is developed by IceWarp Ltd., and can be run on both Windows and Linux. It was formerly known as Merak Mail server.

IceWarp Mail Server consists of a number of modules, with each module representing a particular core feature. However, some modules contain feature subsets.

**Features of IceWarp Mail Server**

**Mail Server**

IceWarp Mail Server, often referred to as IceWarp eMail Server, is incorporated with numerous other modules, including antispam, groupware, and the antivirus modules.

It supports protocols and methods, such as

▸ SMTP/ESMTP
▸ IMAP ACL
▸ IMAP4 with PUSH via IDLE command
▸ POP3/SPOP3
▸ APOP
▸ HTTP(S)
▸ FTP(S) with OTP/S-Key
▸ OpenLDAP
▸ SIP
▸ SIP SIMPLE
▸ OMA DS 1.2,
▸ XMPP
▸ HTTP Proxy
▸ TLS/SSL
▸ IPv6 including AAAA DNS records
▸ SNMPv2
▸ WebDAV
▸ GroupDAVr
▸ CalDAV
▸ SyncML 1.1

It also implements vCard, vCal, iCal,vFreeBusy formats, vNote, quoted/base64 encoding, and Unicode (UTF-8),MD5/SHA1/DigestMD5 RSA encryption methods.

## Web Mail

IceWarp WebMail module has support for the following browsers:

- ▸ Chrome
- ▸ Firefox
- ▸ Internet Explorer
- ▸ Safari

This module can also be integrated with antispam, groupware, and antivirus.

## Web IM

IceWarp Web IM can be integrated with webmail module of IceWarp.

## Antispam

IceWarp Server offers two main techniques for managing spam which arrives at the server:

- ▸ IceWarp Anti Spam Engine - based on conventional antispam methodologies
- ▸ IceWarp Anti Spam LIVE Service - includes real-time antispam protection. The LIVE service uses technologies created by Cyren.

**Antivirus**

IceWarp Server comes equipped with a built-in Kaspersky Anti-Virus Engine.

**GroupWare**

IceWarp GroupWare presents a well-built architecture that allows users to conveniently share information, collaborate and coordinate. The following components are contained in this module:

This module contains the following components:

- ▸ **IceWarp Files** – these allow users to easily share files. The network administrator has the ability to assign accessibility levels for specific individuals as well as groups.
- ▸ **IceWarp Tasks** – these empower supervisors to set assignments for individuals. The assigned tasks may include things such as deadlines together with reminders which are sent by email or appear as pop-ups or instant messages. Appended instruction & deadlines can also be included while monitoring the progress.
- ▸ **IceWarp Contacts**– this acts as a web-based address book. Just like other modules of groupware, the administrator can assign accessibility levels for certain individuals and/or groups.
- ▸ **IceWarp Calendars** – these record deadlines & events. The types of calendar include Personal, Group and Global.
- ▸ **IceWarp Notes** – these allow users to safely document any information.

## 4. IBM Domino

IBM Notes and Domino is a sophisticated platform designed by IBM for hosting social business apps. It is highly scalable, allowing businesses to the flexibility to expand, and features secure applications at a significantly lower cost. It greatly improves productivity and makes decision-making process more accurate and streamlined.

IBM Domino was formerly known as IBM Lotus Domino. It provides many advantages, including the following:

**Reduced cost of ownership** – it reduces the costs of ownership by bringing down the needs of administration by automating not only installations, but also policies and monitoring of tasks, thereby allowing  the staff to focus on bigger things that require more attention.

- It can work on a wide variety of hardware platforms, operating systems as well as directory & client access. Also, depending on your IT budget and overall strategy, you can deploy a wide range of configurations.
- It offers both an administration client and a web browser interface, allowing for a smooth & efficient management of the IBM Domino servers.
- The Domino Domain Monitoring shows – for one or multiple domains – the status of the servers. It becomes easy to identify any issues with the client or the server within a few minutes.
- The Domino Configuration Tuner assists in resolving server misconfigurations, security issues, and performance bottlenecks.
- The Smart Upgrade Technology allows upgradation of the software from a central location, thereby minimizing help desk visits.

**Minimized infrastructure** – it uses less memory, power, network resources, storage, and servers. This is great for companies that do not wish to pitch out additional expenses for running mail servers and wish to do so on their existing IT infrastructure.

- Decreased need for hardware resources, which considerably reduces the demand for large IT budgets, including the amount spent on power, data storage, memory as well as the labor needed for administrative tasks.
- Reduced requirements for network input and output and disk usage through advanced compressions features included in IBM Domino for all documents that contain rich text, images, and graphics.
- Improvement in performance and minimize disk space using the Domino attachment & object service that stores only a single copy of an particular attachment on the server.

**Feature-rich application environment** – it supports people-oriented and workflow-driven applications that assist in a wide array of business applications to enhance the overall performance of the business.

- Deploy third party or custom built applications supporting a wide range of crucial business processes such as customer relationship management, customer support, human resources, sales force automation, billing systems, project management, and much more.
- Utilize the integrated support in workflow engine for applications that make it easy for processes based on spreadsheets or paper documents

**Security-rich** – security is crucial when it comes to a mail server, and IBM Domino provides a security rich environment to guard crucial business assets. It empowers administrators with complete control over information access and protection while guaranteeing document authenticity.

Public-key infrastructure is used to authenticate IBM Notes servers and users. It also verifies digital signatures & encrypts applications and messages, including separate fields within the applications.

# Features Comparison of Microsoft Exchange and IBM Domino

This section will compare the various features of Microsoft Exchange and IBM Domino in a tabulated format.

| Microsoft Exchange | IBM Domino |
|---|---|
| Active Directory (AD) | Notes Address Book (NAB) |
| Local Contacts | Local Contacts |
| Offline Address Book (OAB) | Mobile Directory (MDB) |
| Web Service based Public Folder | NRPC/Replication |
| OAB Generated on mailbox server | MDB Generated on Directory Catalogue (DC) server |
| CAS server copies OAB from MBX server, then sends to user | A replica put into the user's home server |
| Outlook CachedMode>AutoDiscover>CAS | Notes Client (replicatask>dconfig>Home server policy |
| Suggested Contacts | Recent Contacts |
| Auto-Complete | Type Ahead |

Source: http://www.slideshare.net/maria_francis1983/compare-exchange-and-domino-features

# Mail Routing

| Microsoft Exchange | IBM Domino |
|---|---|
| Based on Exchange Attributes in Person's AD Account | Person's PAB information |
| AD site based | NNN based |
| Implicit intra-organization Send connector | Servers having same NNN |
| AD site link for inter-site | Connect docs/NNN for inter |
| Link cost | Connect doc cost |
| Hub role | Notes router |
| Default protocol is SMTP | NRPC for inter-Domino and SMTP for internet |
| Send and receive connectors | Foreign SMTP domain, SMTP connection document and SMTP listener |
| Accepted Domains | Global Domain and Aliases |

Source: http://www.slideshare.net/maria_francis1983/compare-exchange-and-domino-features

# Deleted Items/DB Recovery

| Microsoft Exchange | IBM Domino |
|---|---|
| Deleted Items | Trash |
| Recover Deleted Item | Recover from Backup |
| Purges (Single Item Recovery) | Recover from Backup |
| Litigation Hold/Recover from Backup | Recover from Backup |
| Soft Deleted (Hiding Only)/Disconnected DB | AdminP Approval/Recover from Backup |

Source: http://www.slideshare.net/maria_francis1983/compare-exchange-and-domino-features

## Search

| Microsoft Exchange | IBM Domino |
|---|---|
| Column-wise | Column-wise |
| Advanced search (Property based) | Advanced search (Field based) |
| Exchange Search/Windows Search | Server FT Index/Local FT Search |
| Attachment indexing based on filters | Attachment indexing based on KeyView filters |
| Discovery mailbox searches for all mailboxes in the organization | Domain indexing for file system files/domain databases in a single place |

Source: http://www.slideshare.net/maria_francis1983/compare-exchange-and-domino-features

## User Management

| Microsoft Exchange | IBM Domino |
|---|---|
| AD users, computers, and EMC | Lotus notes admin client |
| EMS | Third party apps/Live console |
| Scripts | Scripts |
| New_requests (Move) | AdminP requests |

Source: http://www.slideshare.net/maria_francis1983/compare-exchange-and-domino-features

## Clients

| Microsoft Exchange | IBM Domino |
|---|---|
| Outlook (MAPI RPC) | Lotus Notes Client (NRPC) |
| OWA (HTTP) | iNotes (HTTP) |
| IMAP | IMAP |
| POP3 | POP3 |
| Active Sync | Traveler |
| BES | BES |
| Web services (built-in) | Customizable |

Source: http://www.slideshare.net/maria_francis1983/compare-exchange-and-domino-features

## Offline Mode

| Microsoft Exchange | IBM Domino |
|---|---|
| Outlook cached mode/offline mode | Offline |
| Local outlook.ost file | Local mail.nsf file |
| Archive.pst | Archive.nsf |
| OAB and local contacts | MobileDir and Local contacts |

Source: http://www.slideshare.net/maria_francis1983/compare-exchange-and-domino-features

## Access Delegation

| Microsoft Exchange | IBM Domino |
|---|---|
| Full access to mailbox (open MBX in Outlook) | Reader access |
| Receive As (log on) to mailbox DB | Reader + Access |
| Send As | Author + Access |
| Folder-wise/Calendar access by owner | Design access/access delegation preferences by owner |

Source: http://www.slideshare.net/maria_francis1983/compare-exchange-and-domino-features

## Archive

| Microsoft Exchange | IBM Domino |
|---|---|
| Personal folder (.pst on PC) | Archive stored on local |
| Personal archive | Archive stored on server |
| Auto archive based on Retention policy settings | Based on archive profile settings and policies |

Source: http://www.slideshare.net/maria_francis1983/compare-exchange-and-domino-features

## High Availability and Fail-Over

| Microsoft Exchange | IBM Domino |
|---|---|
| DAG | Clusters |
| Continuous/Log shipping replication | Event based cluster replication |
| Active Manager | Cluster Manager |
| Mail delivered to active MDB copy | Home mail server copy |

Source: http://www.slideshare.net/maria_francis1983/compare-exchange-and-domino-features

# Reasons to Migrate from IBM Domino to Microsoft Exchange

Microsoft Exchange is an extremely reliable, flexible, and cost-effective email communication and collaboration platform. It consists of some of the most powerful features available in any such types of platforms.

As newer versions of Microsoft Exchange are released, latest features are incorporated making it more and more powerful and efficient.

Let us discuss some of the features of MS Exchange that make it stand out.

## Cloud Network Support

Microsoft Exchange gives you the ability to shift to the cloud. It puts a limit on the number of user disruptions from mailboxes from a wide range of environments, hence providing easier meeting scheduling and calendar sharing no matter what the mailbox environment is.

## Easier Management and Control

The following features have been implemented into Microsoft Exchange to make management and control easier:

Data Loss Prevention (DLP) and mailboxes for the Administrators to allow them to manage directly from Exchange Administration Center (a web-based administration interface)

Role-based access control for people who do not have complete administrative rights, such as specialist and helpdesk users)

## Flexibility and Reliability

MS Exchange offers improved deployment options as well as enhanced flexibility for the administrators to deploy based on the requirements of the organization and its users. It also features numerous hardware storage support options, including that for Storage Area Networks and Direct Attached Storage. A reduced disk input and output per second for the mailbox data as well as protection for mailbox data is also provided.

## Mailbox Archiving Features

The users have the ability to archive mails by keeping certain mails into a primary mailbox; the rest can be placed into In-Place Archive.

## Powerful Retention Policies

Microsoft Exchange offers an enhanced way to meet storage & compliance demands by providing powerful and flexible retention policies.

## Communication across Multiple Devices

Microsoft Exchange allows communication across multiple devices through:

- An intuitive, uncluttered and touch enabled inbox;
- 'Site mailboxes' that enable a user to work efficiently in a team through better collaboration and sharing of information. Exchange also provides document access and messaging along with document storage and co-authoring. SharePoint handles versioning;

- It can also integrate Outlook & Outlook Web App. You can access web-based apps from within Outlook and Outlook Web App. This feature saves time that would otherwise be spent on switching apps – thanks to the Single Sign On access to multiple apps.

## Improved Security and Privacy

- Strong anti-spamming, filtering, and anti-phishing features provide greater protection from email threats;
- The implementation of powerful internal compliance policies provide better authorization by preventing sensitive information of users from being accessed by unauthorized personnel. The Data Loss Prevention (DLP) is a great example of such features – it prevents violation of any policy or access without proper authorization. Working on the basis of PCI and PII, the DLP also adheres to numerous other authorization policies as well as regulatory standards;
- A single interface allows the user to work in compliance with numerous different environments, such as Exchange, Lync, and SharePoint.

There is also support for industry standard protocols. These protocols provide 'Anywhere Access' to the data in a mailbox apart from providing other email features, including instant messaging and voice mail.

## Exchange versus Domino

Microsoft Exchange is a commonly used platform due to its broad features, such as emailing, contacts management, and calendaring on various platforms such as mobile, computer, or web. This enables users to connect with others and enables efficient data synchronization.

On the other hand, IBM Notes and Domino features numerous built-in collaboration functionalities, just like Exchange, including calendaring, emailing, and contacts management and instant messaging,

So what is it that makes Exchange stand out? Let's find out:

- Exchange offers the user the ability to create a whole unified messaging solution. On the other hand, when it comes to Domino, some additional products from IBM have to be installed to set up a complete unified messaging solution. Therefore, Exchange wins over Domino in this regard;
- Exchange also allows integration of voicemail into Outlook Web App. SATA hard drives are supported. IBM Notes and Domino do offer some features such as the likes of auto email address completion and mashing up of the calendars, as well as the support for, and receiving of, vCards. Despite of these new features, it still tends to lacks the powerful feel of MS Exchange;
- Microsoft Windows operating systems are the most common OS used for business and personal use, mainly due to their extremely easy to use GUI. This is yet another factor that makes MS Exchange so popular: it is undoubtedly the most compatible mail server for Windows thanks to its superb support for Windows-based systems. The entire installation process is easy and bug-free;
- When it comes to cost, Exchange offers greater flexibility and more control over different aspects of IT than IBM's Domino. It also offers more deployment options than the latter;

- A native storing and archiving ability along with retention capabilities make Exchange superior to Domino's limited features, not to mention the additional complexity and costs involved in enhancing the latter platform. Microsoft has implemented support for various types of storage hardware, including SATA, DAS, and JBOD. While Domino does have the Domino Attachment Object Store

(DAOS), it is still not as efficient when compared to that of Microsoft Exchange;

- With every release, Microsoft not only adds additional powerful features, but also pays great attention to the software's performance. Starting from the earliest versions, the performance of Exchange has improved dramatically with each new release. For instance, in Exchange 2010, there is a 70 percent improvement in performance as compared to that of the 2007 version. Unfortunately, such great performance improvements have not been seen in IBM's releases;
- Exchange offers highly innovative tools as well as capabilities for the entire global economy. For example, it has a calendar sharing ability allowing users within an organization to share crucial data with users outside, There are also numerous permissions and securities in place that prevent any unauthorised access to the data;
- The built-in enterprise mobility features of Exchange give users a truly universal inbox. This 'universal inbox' saves the user's mail data, voice mail, SMS, and IM conversations in a single place, allowing for easy retrieval. Network administrators possess rights to make decisions regarding which devices are permitted and which are restricted to access the network. Everything comes built-in with Exchange. Mobile data security is taken care of through implementation of numerous data protection policies. On the contrary, Exchange's counterpart only offers a bare minimum and basic mobile experience to its users.
- Exchange enhances the overall user experience by providing:
  - An improved mail conversation view;
  - Speech-to-text transcription for the voice mails;
  - Enhanced Messaging & restrictions for unwanted; emails through MailTips. This improves messaging efficiency and avoids unintended or accidental mails.

Based on the information above, it would be safe to say that Exchange is indeed a more powerful and flexible option. While the above comparison has revealed a few differences between the two applications, it should not be ignored that both have their own sets of advantages and disadvantages. One piece of software may be suitable for someone while the same may not meet the needs of the other.

With that said, it is crucial that you analyze your own needs and requirements before you make a choice.

Once your needs have been clearly established, a detailed examination of each of the feature set will reveal which option is better for you.

**For more information on Microsoft Exchange, visit https://products.office.com/en-us/exchange/email.**

**For more information on IBM Domino, visit http://www-03.ibm.com/software/products/en/ibmdomino.**

# Open Source

So far, we have mentioned a few commercial mail server products. In this section, we will highlight one of the most common open-source mail server options.

Open source software allows a user to study, modify, and distribute code to anyone for any purpose as this software is distributed a collaborative, public manner.

Unlike commercial products, which often come with a hefty price tag, open source mail server products are freely available. Their sophistication and feature set vary significantly, however, and in some cases, they may not be able to compete with the extensive nature of some commercial products available in the market.

## 1. Postfix

Postfix is an open-source, and free, mail transfer agent (MTA) designed to route and deliver email. It is released under IBM Public License 1.0.

It was originally developed at IBM's Thomas J. Watson Research Center by Wietse Venema and released to the public in 1998. It continues to be actively developed by the creator and numerous other contributors to date. According to a research conducted in 2013, it was found that as many as 27 percent of all public mail servers available on the Internet had Postfix running.

Postfix, as an SMTP server, provides the first layer of production against malware and spambots. Server administrators can merge Postfix with various other types of virus/spam filtering, message store access, or sophisticated SMTP level access policies.

Postfix also features a high performance mail delivery engine, and is often combined with mailing list applications.

## Features of Postfix

- Standards-compliant support for SMTP, SMTPUTF8, LMTP and STARTTLS encryption inclusive of the DANE protocol support & "perfect" forward secrecy, MIME encapsulation and transformation, SASL authentication, IPv4, and IPv6, and DSN delivery status notifications.
- Configurable SMTP-level access policy which can automatically adapt to overload
- "Virtual" domains along with distinct address-namespaces
- UNIX-system interfaces for delivery to command, command-line submission, and for direct delivery to the message stores in maildir and mbox format.
- Light-weight content inspection on regular expressions
- Various database lookup mechanisms such as CDB, Berkeley DB, Memcached, OpenLDAP, LDAP, LMDB, and various SQL database implementations.
- A refined scheduler which can implement parallel deliveries together with back-off strategies & configurable concurrency
- A scalable zombie blocker which can reduce SMTP server load as a result of a botnet spam

# Chapter 4: Email Clients

An email client, formally known as a mail user agent (MUA) is a type of software that is used to access a user's e-mail account. There are many types of e-mail clients available - both free and commercial. Due to the fact that e-mail has been around for a few decades now, the majority of companies have continuously improved their versions of e-mail clients and added a lot of features; however, the basic working of each of these clients remains the same.

E-mail clients are locally installed on a computer. There are, nonetheless, web applications that provide the same type of features as e-mail clients (and are also considered to be one), but are more commonly called *webmail.* These web-based e-email clients include the highly popular Gmail, Outlook, and Yahoo! Mail. These e-mail services will be discussed in greater detail in Chapter 7; for now, our focus will be on local e-mail clients.

## Receiving Mail from an E-Mail Account

Like the majority of client programs, an email client only becomes active when it is manually run by a user. The most common way for the email client to work is to make an arrangement with a Mail Transfer Agent (MTA) server for the reception of, and storage of the user's e-mails.

It is the MTA which, through a Mail Delivery Agent (MDA), adds the e-mail messages to the user's mail storage as soon as they are received. The user's storage remotely located on the server is known as the mailbox.

On many UNIX systems, the default setting for a mail server is to store messages that have been formatted in *mbox,* inside the HOME directory of the user.

These e-mails are stored in the mailbox of the user until he or she decides to retrieve them using an e-mail client, and download them onto their computer. The same e-mail client can be configured to access and retrieve e-mail from a number of remote servers either on-demand or at a set-interval.

An e-mail client can access this mailbox in two different ways:

1. **Post Office Protocol (POP)**
2. **Internet Message Access Protocol (IMAP)**

**Post Office Protocol** or **POP** allows the mail to be downloaded individually. These e-mails are deleted from the server once they have been successfully downloaded and saved on the local computer. There is an option to leave a copy of the message on the server to enable another client to access it, if needed. However, there is no option that allows flagging of individual messages, such as 'seen', 'answered', or 'forwarded'. With that said, POP is not the right way to access mail for people who use different computers to do so.

The **Internet Message Access Protocol** or **IMAP**, on the other hand, enables users to keep the messages on the remote server, allows flagging, and deletion of messages without having to download them on local storage first.

IMAP also provides folders along with sub-folders which can be easily shared with different users. Different access rights may also be assigned. This is an excellent option for organizations that need to have multiple employees access the same e-mail account but with varying privileges.

By default, folders such as Drafts, Sent, and Trash are created. IMAP also includes an idle extension feature for faster notification and real time updates.

Also, access to mailbox storage can be gained directly by applications that run on the server, or through shared disks.

While direct access can be effectual, but it is less portable because it depends on mailbox format, it is utilized by some e-mail clients along with certain webmail applications.

## Composition of Messages

Email clients come with a user interface that allows the users to view and edit text. Some email clients out there also permit use of an external editor for composition of messages.

For the headers and the body, email clients have to carry out formatting based on the specifics mentioned in RFC 5322. For non-textual content as well as the attachments, MIME is the standard. The headers consist of the destination fields, such as the *To, Cc,* and the *Bcc,* as well as *From* – the originator field that mentions the author of the message. If there are more than one authors, then the *Sender* field and *Reply-To* if the reply is to be directed to a different mail account.

To help users with destination fields, a number of clients feature address book(s) or have the ability to connect to a certain LDAP directory server. Different identities for originator fields may also be supported.

The client settings normally require the real name of the user and their email address for each identity of the user, and in some cases, a list of the LDAP servers.

## Sending Messages to the Server

The email client handles everything ranging from message composition to sending the email to the server. The client is normally set up to establish a connection with the user's email server – one that is normally an MTA or an MSA – which are two variants of the SMTP protocol. An 'authentication extension' is created by the email client using the SMTP protocol, this authentication extension is used by the mail server to authenticate the sender of the email. The technique is used because it simplifies nomadic and modularity computing.

Prior to this method, the mail server used to recognize the IP address of the client. For example, if the client was present on the same computer, it used 127.0.0.1 as the internal address – or because of the fact that the same ISP provided not only the internet services, but also the email services.

The name and the IP address of the outgoing mail server are required in the client settings, including the user name and password for authentication purposes, and the port number.

Port 25 is for MTA, port 587 is for MSA, and for backward compatibility, the majority of clients and servers have support for the non-standard SSL port 465 for SMTP.

# Encryption

Without encryption, an email message would be no different from a postcard, open for anyone who wishes to eavesdrop. Email encryption ensures privacy of the email by encrypting not only the body of the email, but also the mail sessions. Without encryption, any individual with network access, the right tools, and skills could monitor the content of emails exposing possible sensitive information.

## Mail Session Encryption

All of the protocols used for email have an inherent option to allow users to encrypt the entire session. This prevents the user's login credentials from being 'sniffed'. These protocols are recommended to be used whenever the ISP is not to be trusted.

Whenever sending email, users are only able to control the encryption right from the very first hop to the server from the client. If any additional hop is performed, the messages may or may not be transmitted with encryption. This would depend on the capabilities of both the transmitting server and the one receiving it.

## Message Body Encryption

There are two different methods for managing the cryptographic keys involved in encryption of the body of email messages. S/MIME is based upon a model that is dependent on the use of a trusted certificate authority (CA) to sign the public key of users.

OpenPGP, on the other hand, is somewhat a flexible mechanism as it enables users to sign the public key of one another. It is also flexible when it comes to the message's format in the sense that it has support for plain email encryption as well as signing just like they worked before the standardization of MIME.

In both the cases, the message body is the only portion that is encrypted. The header fields, such as the information about the originator, the recipient, and the subject are transmitted in plain text format.

## Webmail

Email clients are not limited to the applications that are downloaded and installed locally onto your computer. Web mail services also have an email client of their own that runs on their server – allowing the user to login to their accounts from their website's email portal. Webmail offers numerous advantages, including the capability of sending and receiving email from a web browser regardless of which computer is being used to access it (as compared to locally installed email clients that require configuration on individual computers).

We will discuss email services in detail later on in this book, however, for now, it is sufficient to know that popular email servers such as Outlook.com, Gmail, and Yahoo provide email services with web-based email clients for the convenience of users, along with giving an option of using protocols such as POP and IMAP for accessing the mailbox.

# Protocols

The most popular protocols used for retrieving emails include IMAP4 and POP3, and SMTP is used for sending emails.

The majority of email clients support MIME, which is an important standard using to transmit binary file mail attachments. Attachments are those files that are actually not a part of an email, but are 'attached' to it to be sent along with them. Most of the email clients use a 'User-Agent' header to identify the application that has been used to send the email.

## Port Numbers

Email clients and servers use certain TCP port numbers by convention. These are depicted in the following table:

| protocol | use | plain text or encrypt sessions | plain text sessions only | encrypt sessions only |
|---|---|---|---|---|
| POP3 | incoming mail | 110 _pop3._tcp | | 995 _pop3s._tcp |
| IMAP4 | incoming mail | 143 _imap._tcp | | 993 _imaps._tcp |
| SMTP | outgoing mail | 25 | | 465 |
| MSA | outgoing mail | 587 _submission._tcp | | |
| HTTP | webmail | | 80 | 443 |

SOURCE: Wikipedia

# Most Popular Email Clients

The following email clients are extremely popular. Equipped with latest features and support for various features, including multiple protocols, spam control, and others, they meet the needs of even the most demanding email users.

## Microsoft Outlook

Outlook has been in existence since as back as the early 1990s. As it comes in the Microsoft Office bundle, it has been deeply ingrained in corporate environments, and is immensely popular. Just because it comes bundled with Microsoft Office does not imply that people are 'forced' to use it. Outlook is renowned for its merits. For instance, it seamlessly integrates with the Windows Desktop Search, thereby allowing users to search through their entire Outlook workflow.

Additionally, Outlook handles everything from email to tasks, contacts, and calendar, making it a comprehensive 'organizer' for personal and professional use.

### Essential Tips for Beginners

Outlook 2013 is an incredibly powerful software that, when used properly, can save you a lot of time and enhance your productivity levels. Here are a few tips that will help you get to know Outlook better.

▸ **Leverage Outlook's Advanced Search Features**

Clicking on the Search field on the top of the message list reveals 'Search tools' – a new tab containing numerous search options. In this tab, click on the down arrow next to 'Search tools', and click on Advanced Find. This will open a dialogue box that will allow you to fine-tune your search results.

▸ **Fine Tuning Clean Up Options**

Outlook 2013 comes with a Clean Up Conversation feature. You can further fine-tune it to make it work as per your needs. Click on Settings instead of clicking on the Clean Up button to reveal an appropriate section of Outlook Options.

You will find a section that enables you to create a folder or specify one for storing the messages that have been deleted by Clean Up until you decide to permanently delete them. If no folder is specified, then the messages deleted by the Clean Up feature go directly to the Deleted Items folder.

These options can be accessed anytime by navigating to File → Options → Mail.

### ▸ Auto-Close Outlook Message Window

After replying to a message, you surely don't wish to keep staring at the email. You can configure Outlook to automatically close the message after the reply has been sent. Navigate to File → Options → Mail and scroll down to the portion where it says *Replies and Forwards.* Place a checkmark next to the option that says 'Close original message window when replying or forwarding."

Now that you are already in this section, do not click OK just yet. Proceed to the next tip as it has to be configured in the same section.

### ▸ Utilise Reply and Forwarding Features

You can instruct Outlook what to do with the original message when you are replying or forwarding it. Navigate to File → Options → Mail (if you aren't already there), and then move down to the Replies and Forwards section.

Choose from any of the following options as per your requirements:

- o  'When replying to a message', and
- o  'When forwarding a message'

### ▸ **Resend or Recall Outlook Message**

Prior versions of Outlook had a feature that allowed recall or replacement of an already-sent email message if both the sender and the recipient were connected to a particular MS Exchange Server.

This feature has been extended by Outlook 2013 to include any recipient that uses Outlook. This ability, however, depends on the settings of the recipient's Outlook. Nevertheless, it is worth trying if you have sent out a message containing incorrect information or information that should not have been sent in the first place.

To make use of this feature, open up a message in your Sent items folder, click on File → Info → Message Resend and Recall, and then follow the instructions. Outlook will inform you whether the process has succeeded or failed.

### ▸ **Empty Trash on Exit**

If you are fed up of hundreds of email messages piling up in your trash folder, then you should make use of Outlook's empty trash on exit feature. Click on File → Options → Advanced, and get to the Outlook Start and Exit section. Select the checkbox right next to the *Empty Deleted Items folder when exiting Outlook*.

### ▸ **Outlook Mobile options**

Large corporations can benefit from Outlook Mobile options to send out calendars, reminders, and messages directly to a mobile device.

However, this is not limited to corporations as individuals can also utilize this feature. Head to File → Options → Mobile and choose the relevant options.

A menu appears that allows you to choose an SMS service provider. You can enjoy free trials from some providers with no credit card details required at all.

You can configure the service to connect to your smartphone, and you will be provided with clear instructions for setting up the SMS account in MS Outlook.

### ▶ Categorize by Colour

Many advanced features of MS Outlook, such as the one discussed above (mobile options), utilise the categories that you have designated to tasks and messages. By default, such categories are named after the colour that appears right next to the message in a box. To make things more effectual, you should name the categories under more meaningful names, such as 'Weekend' or 'Urgent' or something that clearly portrays the purpose of the category.

Go to the Home tab in Tags group, click on Categorize, and then click 'All Categories' to rename and to select the required colour.

### ▶ Flag Outgoing Messages

When composing a message, you can set up a reminder to alert you to follow up on the same message later on. While you have the message editor window in front of you, open the Ribbon's Message tab. In the Tags group, click on Follow Up, and you will get a dropdown box menu where you can select the time for the follow up to be displayed. If you click on Custom, the menu displayed will allow you to fine-tune all other options for the reminder.

### ▶ Add Business Card

You can easily add business cards In Outlook as a signature that will allow your recipients to add all your contact details to their address book. Proceed to File → Options → Mail, and click on Signatures.

In the Email Signatures tab, click on New and select a convenient name for it. Next, click on Edit Signature pane at the bottom of the tab and select Business Card. You can now choose your own contact information from the Outlook's contact list.

As a final step, assign the signature in the top right area of the tab as the default for all your outgoing messages.

## Apple Mail

Apple has developed the Mail application for use on the Macintosh platform. It follows Apple's 'it just works' methodology and enables users to easily gather email from one or more servers conveniently. It is an intelligent piece of software and engages with you. For example, if you receive an email with an invitation of some kind for a particular date, it automatically adds it in the calendar – reminding you when the time comes. Just like Outlook integrates with Windows Desktop Search, Apple Mail application can also integrate with Spotlight to make messaging easier.

## Mozilla Thunderbird

Thunderbird is offered by Mozilla as an open-source application. It is a solid email that has support for extensibility, just like Mozilla's Firefox web-browser. You can easily add extensions to enhance Thunderbird's features. A portable version of Thunderbird is also available that can be carried around on a USB drive and can be used to access email when on the move.

## Postbox

Postbox is an email client that is available for both Windows and Macintosh systems. Due to the fact that it is based on Mozilla's code, the team at Postbox has been successful in tweaking many Thunderbird extensions to work on Postbox. The interface is quite powerful and includes features that allow you to compose and search for email simultaneously – you can compose an email while searching for a previous mail in the sidebar. It also displays summaries of the email as you search, including key information such as numbers or links from inside the email, instead of just the Sender name and the subject.

# Chapter 5: DNS Setup for Email

Imagine what would happen if you had to go somewhere and you had no address for the destination. Would you be able to reach your destination? Of course not. Similarly, your web browser accesses websites based on unique IP addresses. You do not necessarily enter these IP addresses; in fact, you may enter something such as 'xyzwebsite.com' in your browser, and the website opens up.

This is where DNS comes in. DNS can be likened to an address book. Upon entering a website address, the browser looks up for its appropriate IP address through DNS, hence retrieving it from a huge database. Sounds all too simple? Let's get into the details of DNS and what role it plays in email.

## Understanding DNS

Domain Name System or DNS in short, is among one of the most crucial components that make up the Internet infrastructure. If there is no DNS available, you would have major problems in finding the required resources on the net; equally, others will also be powerless to locate you. All of this is because DNS is an address book (or phone book, call it what you like) that translates website addresses such as www.google.com into IP addresses such as 74.125.224.72, and of course vice versa. DNS eliminates the need for us to memorize IP addresses of the websites we frequent.

Being able to find hosts by name enables IP addresses to be changed with time as websites grow and change their location – or if they reconfigure. However, DNS offers a lot more services than name to address mapping. It is crucial for network administrators to grasp the basic structure, operations, and function of DNS.

We will begin by discussing DNS in general before proceeding towards actually setting up DNS for email.

**DNS** is a distributed database, is hierarchical and with a delegated authority. The terms 'delegated authority' implies that you are liable to provide a way for users on the Internet to find IP addresses that are linked with your company's domain. Some businesses let their Internet Service Providers (ISPs) take care of DNS; however, this is not always a wise idea. A slight mistake during configuration by the ISP can make a company's website appear offline. You may also lose the control you have over your domain's information.

DNS is the key to your online existence – this is precisely why you should control the DNS for your website's domain. DNS also incorporates an anti-phishing mechanism, allowing companies to reject spam email. On the other hand, it also acts as a privacy mechanism, which effectively conceals the internal network topology.

DNS helps in the following areas:

**Anti-Phishing** – When a DNS is configured properly and is doing its job, it helps you to reach the correct website and not an imitation one being run by identity thieves.

**Anti-Spam** – You would have been receiving a lot more spam than you are now had it not been for DNS. The mail server can be configured to verify the domain name of incoming email, thereby filtering spam. More will be discussed on DNS and its relation with email servers. Various new mechanisms, such as Sender Policy Framework (SPF) and the DomainKeys (DKIM) can be used to identify the allowed senders and to reject email from unauthorized ones. Additionally, real-time blacklists (RBL) enable email servers to check whether a sender is in the spam list or not.

**Privacy** – DNS only reveals information to clients that you want the public to be able to see with regards to your network. Similarly, it also shows internal users and the servers what is appropriate for them.

## DNS Records

The ultimate purpose of the DNS is to map the host names to their appropriate IP addresses. The data that enables this to happen is stored as *records* on a DNS server's zone file. Included inside each zone file are the resource records, which define not only the host names, but also numerous other elements of domains.

The following table lists the types of resource records that are supported by Windows 2000 DNS service:

| Record | Purpose |
| --- | --- |
| SOA | Specifies the authoritative server for a zone |
| NS | Specifies the address of a domain name server or servers |
| A | Maps the host name to a particular address |
| PTR | Maps the address to the host name for purpose of reverse lookup |
| CNAME | Creates an alias (synonymous) name for a specified host |
| MX | The mail exchange server for a domain |
| SRV | Defines the servers for a specific purpose such as FTP, HTTP, etc. |
| AAAA | Maps the host name to an IPv6 address |
| AFSDB | Location of an AFS cell database server or the DCE cell's authenticated server |

| HINFO | Identifies the host's hardware and the OS type |
|---|---|
| ISDN | Maps the host name to the ISDN address (phone number) |
| MB | Associates the host with a specified mailbox; (experimental) |
| MG | Associates the host name with a mail group; (experimental) |
| MINFO | Specifies a mailbox name responsible for a mail group; (experimental) |
| MR | Specifies a mailbox name that is proper rename of the other mailbox; (experimental) |
| RP | Identifies THE responsible person for a domain or a host |
| RT | Specifies intermediate host that routes the packets to a destination host |
| TXT | Associates the textual information with an item in the zone |
| WKS | Describes the services provided by a specific protocol on a specific port |
| X.25 | Maps the host name to a X.121 address (for X.25 networks); used in combination with the RT records |
| WINS | Allows lookup of the host portion of a domain name through the WINS server |
| WINS-R | Reverses the lookup through a WINS server |
| ATMA | Maps the domain name to an ATM address |

While there are many records within DNS, we will now talk about MX records as it has everything to do with a mail server.

## Mail Exchange Record (MX Record)

MX records are utilized by mail servers to figure out where the mail should be delivered, and these records are only mapped to A records. If an MX record is not present for a domain, the delivery of the mail will then be attempted by matching with 'A' records. Thus, if a domain called 'xyzdomain.com' does not have an MX record, the mail would be tried to be delivered to the root record of 'xyzdomain.com'.

### Where are the MX records?

Your MX records would be on the server of the DNS provider. If you change your MX record that are with your DNS provider, all other servers will also copy the updated versions of the MX records.

### How can you see your MX records?

To find your latest MX records, follow the steps given below:

1. Visit http://mxtoolbox.com/

2. Enter your domain name in the box, click 'MX Lookup'

3. You will be presented with all the details about your MX records

### What does an MX record contain?

An MX record contains the following fields of information:

**Name** – This will be the name of your domain

**Class** – The Class is always set to 'IN', meaning Internet.

**Type** – This is set to MX for MX records

**TTL** – Time to Live. This indicates the time it will take for an MX record to update. This time is measured in seconds; for instance, a TTL of 1800 will mean half an hour to update the database. A higher TTL implies less load on DNS server; however, it also implies that it will take longer to change the MX records.

**Priority or Preference** – This will be the order of preference for delivery of mail. The sending servers ought to try the lowest number of preference first, followed by the next lowest, and the like.

**Data** – This will be the host name of the server that is responsible for handling mail for the specific domain.

For example, if your domain is xyzdomain.com, the MX records for it may appear as follows:

---

**xyzdomain.com. IN MX 86400 1**
**xyzdomain.com.s7a1.psmtp.com.**

**xyzdomain.com. IN MX 86400 2**
**xyzdomain.com.s7a2.psmtp.com.**

**xyzdomain.com IN MX 86400 3**
**xyzdomain.com.s7b1.psmtp.com.**

**xyzdomain.com. IN MX 86400 4**
**xyzdomain.com.s7b2.psmtp.com.**

---

Each DNS host has a unique user interface for the MX records. For instance, some use trailing periods while some do not. On the other hand, some also give you the option of setting the TTL, while some do not.

# Chapter 6: Attachments

'Email attachment' refers to a computer file that is sent along with the message in an email. An email allows attachment of one or more files, depending on what has been allowed by the mail server. Sending attachments through emails is a common way to share files, including, but not limited to, images and documents.

## Size Limitations

Even though some email standards (such as MIME) do not limit the size of an attachment, email users normally notice in practical usage that they are not able to send very large files. This has to do with the way email is channeled through the internet.

An email passes through a number of different mail transfer agents before it successfully reaches the intended recipient. Each time the email arrives at a mail transfer agent, it has to be stored before being forwarded to the next one. This necessitates having an email attachment size limitation. This is yet another reason why large attachments can be easily sent within an organization (internally), but they tend to be unreliable when being sent across different MTAs over the Internet.

Google Mail (Gmail) restricts the file size to 25 MB, stating the following:

*"you may not be able to send larger attachments to contacts who use other email services with smaller attachment limits"*
**SOURCE: Wikipedia**

A message size of 10 MB is considered to be safe. These limitations can puzzle email users due to the fact that MIME encoding utilizes Base64 – which adds around 33 percent overhead, meaning that a document of 20 MB on disk can exceed the message size limit of 20 MB.

For a smoother, safer and easier way to send and receive email attachments, numerous services have been launched with the promise of solving these issues. For instance, eParcel attachment service moves the file attachments from email systems and timestamps each of the 'eParcels' for self-destruction once the file has been downloaded. Using this technique, the mailbox size and email attachment limits can be easily circumvented for both the sender and the recipient.

## Potentially Dangerous File Types

Email users have always been warned and it is a general understanding among regular email users that any unexpected email messages with attachments should be considered suspicious and potentially dangerous. This is particularly true when the email is not from a trusted source.

Nevertheless, this is not enough to counter the prevalent threat of executables programs that are sent by those willing to spread mischief. For instance, as early as the 1987, the mainframe-based file called Christmas Tree EXEC, the ILOVEYOU and the Anna Kournikova worms of 2000/2011 have led to the addition of extra layers of protection to prevent malware infections. With that said, many email providers block specific file extensions altogether. Gmail is one such service that blocks executable files (.exe).

## The Technical Aspects

Initially, the SMTP email was of 7-bit ASCII text only. Files were attached manually by encoding 8-bit files using BinHex, uuencode, or xxencode, and then pasting the results into the message's body.

However, contemporary email systems adhere to the MIME standard, which has made attachment of files almost a seamless act. This method was developed by a person named Nathaniel Borenstein and his collaborator Ned Freed, and the first MIME email with attachment sent on March 11th 1992.

MIME allows the message and any attachments to be directly encapsulated within a single, multipart message. Base64 encoding is used to convert the binary into a 7-bit ASCII, However, on contemporary email services that run Extended SMTP, a full 8-bit support is provided by the 8BITMIME extension.

# Chapter 7: Email Services

Various email services are available that provide their own webmail interface along with support for numerous protocols that allow users to retrieve email from the server using a mail client. These email services have grown rapidly in numbers, with each offering its own set of features and limitations.

In this chapter, we will provide an overview of the most common email services – both free and paid. Some of the services mentioned in chapter are relatively old in terms of the time they have been around for. Most of these have been created or acquired by IT giants, while some are relatively new companies that have ventured into the market and have succeeded in providing top-quality email services.

## 1. Google Mail (Gmail)

Google, the internet giant, started offering Gmail back in 2004 as an invitation-only beta release service. The email service was made available to the public in 2007.

Initially, Gmail offered a storage space of 1GB. This size was a huge increase from what the-then competitors offered. Hotmail, for instance, offered 4MB of space. Gmail is an ad-supported, free email service that supports a message size of up to 25 MB per email. With 425 million users all over the world, it was the most popular email service as of June 2012.

Gmail offers access to the mail through its web-based interface, and via IMAP4 or POP3.

To learn more about Gmail's features, visit www.gmail.com

## 2. Outlook.com

Outlook.com is an email service run by none other than the famous Microsoft Corporation. It is among the world's first web-based email service, having been founded as Hotmail in 1996. In 1997, Microsoft acquired Hotmail for an estimated value of $400 million, re-launching it as MSN Hotmail, followed by Windows Live Hotmail, and now Outlook.

As of 18th February 2013, Outlook had 420 million users. Outlook provides unlimited storage with a maximum limit of 25 MB per message. There is no limitation of file size when attached through OneDrive. Outlook also allows users to access their mail through the web-based interface, through POP3 or IMAP4.

To learn more about Outlook, visit www.outlook.com

## Configuring Advanced Settings for Gmail

Google provides immense flexibility to Google Apps administrators and allows them to configure various email settings for their own domains and for the different groups of users. Before you can start configuring the settings, you need to be aware of the following:

- Prior to customizing the email settings for each group of users, 'organizational units' have to be added while an organizational structure has to be created.

  To learn how to add an organizational unit, visit https://support.google.com/a/answer/answer.py?answer=182537.

  To learn how to create an organizational structure, visit https://support.google.com/a/answer/answer.py?answer=182433.
- It takes around an hour after configuring email settings for them to spread to the accounts of individual users.

- In some rare instances, a few users might experience delays in messages if you configure a lot of Gmail's advanced settings. Such delays would only affect those messages that have a huge number of recipients. The first recipient is ALWAYS accepted regardless of the number of settings configured.

Besides spam and virus protection, the advanced settings in Gmail allow administrators to customize numerous security settings for email for each organizational unit. Some examples of these settings are as follows:

- **Content Compliance** – specify the actions to perform for messages depending on pre-defined words, text patterns, phrases, and numerical patterns.
- **Objectionable Content** – specify the action to perform based on word lists you created
- **Attachment Compliance** – specify the action to be performed for messages containing attachments
- **Append Footer** – configure the outbound messages with a footer text for legal/promotional/informational requirements
- **Approved Senders** – create approved sender list that bypasses spam folder
- **Blocked Senders** – Block certain senders based on their domain or email address
- **Mail Routing and Delivery** – routing and delivery options for dual delivery, split delivery, sending routing, receiving routing, attachment routing, content routing, and more.
- Non-Gmail Mailbox Routing and Quarantine Summary – reroute messages to the users' non-Gmail email accounts. Also, set Quarantine summary reports.
- **Restrict Delivery** – Restrict email addresses the users can exchange email with.

# Gmail Advanced Settings

The Gmail advanced settings enables an administrator to configure numerous settings for Google Apps domains. It also allows configuration of settings for certain organizational units (or groups of users).

Before you can start to tweak the settings, it is necessary to add organizational units and to create an organizational structure. Also, keep in mind that it may take as much as an hour for the new configuration settings to be applied to the individual user accounts.

As mentioned previously, some of your users may experience some kind of a delay in messages, particularly if you have made some major changes to the settings.

## How to Configure Advanced Settings

**Step 1:** Sign in to Google's Admin console

**Step 2:** On the Dashboard, navigate to Apps → Google Apps → Gmail → Advanced settings

**Step 3:** In the *Organizations* area, select the top-level organization or the sub-organization that you are interested in configuring.

**Step 4:** Scroll down to the sections that you wish to configure. The *Search settings* box also allows you to enter a term that will enable you to quickly find the setting that you want to configure.

**Step 5:** Once you have made the changes, click on *Save Changes* to save the new configuration.

## Email Settings Descriptions

The Gmail advanced settings have been divided into six distinct sections, as depicted below. In this section, we will look into each of the sections in detail.

1) **Setup**
2) **End user settings**
3) **End user access**
4) **Spam**
5) **Compliance**
6) **Routing**

Source: https://support.google.com/a/answer/2786758?hl=en

### *Setup*

▶ **Web address** - Change URL of the users' Gmail login page. *(top-level org only)*
▶ **MX records** - View MX records. *(top-level org only)*
▶ **User email uploads** – Enables users to be able to upload the mail using Google's Email Migration API. *(top-level org only)*
▶ **Uninstall service** - See how to disable your email service for more information. *(top-level org only)*

### *End user settings*

▶ **Themes** - Specify whether the users are able to choose their themes.
▶ **Email Read Receipts** - Specify whether or not the users can request/return read receipts (check out the Enable read receipts page).
▶ **Mail delegation** – Enables users to delegate access to the mailbox to other users on the domain.
▶ **Emailing profiles** - Specify if your users can send and receive mails through their Google+ profiles. The default setting is set to OFF.
   o *Note: The setting will only appear if you already have Google+ enabled for your domain.*

- **Name format** - Configure name format for your users, or empower them to customize the setting.
- **Apps search** - Include the relevant Drive documents & Sites in Mail's search results. The search results for apps appear below the search results of mail.

## *End user access*

- **POP and IMAP access** - Disable IMAP and POP access for your users (for more information, visit: IMAP and POP access).
- **Outlook & BlackBerry Support** - Enable Google Apps Connector and Google Apps Sync for your users.
- **Automatic forwarding** - Specify if your users can forward incoming mails to some other address (for more information, visit Disable automatic forwarding).
- **Offline Gmail** - Enable offline Gmail for your users.
- **Allow per-user outbound gateways** – Allow your users to send out mail through external SMTP server.
- **Image URL proxy whitelist** - Create & maintain a whitelist of all internal URLs that are allowed to bypass proxy protection. (top-level org only)

## *Spam*

- **Email whitelist** – Develop an email whitelist (this is a list containing IP addresses deemed legitimate). (top-level org only)
- **Inbound gateway** - Enter the inbound mail gateway if available. (top-level org only)
- **Spam** - Create a list of approved senders in order to bypass spam folder.
- **Blocked senders** - Block certain senders depending on their email address or the domain.

## *Compliance*

- ▸ **Email retention** - Control amount of email stored for each of users. (top-level org only)
- ▸ **Append footer** - Configure the outbound messages with footer text for legal compliance purposes, or for promotional or informational requirements.
- ▸ **Comprehensive mail storage** – Make sure a copy of all received or sent email, including the mail sent/received by the non-Gmail mailboxes, is being stored in respective Gmail mailboxes.
- ▸ **Restrict delivery** - Restrict the email addresses with whom the users can exchange email with.
- ▸ **Content compliance** - Specify the action that needs to be performed for messages depending on certain predefined sets of phrases, words, text patterns, or any numerical patterns.
- ▸ **Objectionable content** - Specify the action that should be performed for messages depending on the word lists created by you.
- ▸ **Attachment compliance** - Specify the action that should be performed for messages with the attachments.
- ▸ **Secure transport (TLS) compliance** – Select the mail to be sent/received through a secure connection, especially when the users correspond with certain email addresses and domains.

## *Routing*

- ▸ **Email routing** – This option offers a set of legacy routing controls for your domains. To get the guidelines and the best practices for configuring routing controls, visit Managing mail routing and delivery. (top-level org only)
- ▸ **Outbound gateway** - Set the outbound mail gateway, a server that allows passage of all the email being sent from your domains. (top-level org only)
- ▸ **Recipient address map** - Apply one-to-one aliases (mapping) to the recipient addresses on all messages received at your domain. (top-level org only)

- ▸ **Receiving routing** - Set up the inbound and the internal-receiving delivery options, including dual delivery & split delivery.
- ▸ **Sending routing** - Set up the outbound and the internal-sending delivery options.
- ▸ **Vault settings for Exchange Journals** - Specify a particular email address within your domain that will receive all your Exchange journal messages. (top-level org only)
- ▸ **Non-Gmail mailbox** - Reroute the messages to your users' non-Gmail mailboxes. This is important for those organizations that use non-Gmail mail services, including MS Exchange or some other non-Google SMTP.
- ▸ **SMTP relay service** - Set the options for routing your outbound mail within Google. (top-level org only)
- ▸ **Alternate secure route** - Set the alternate secure route if the secure transport (TLS) is needed.

## 3. Yahoo! Mail

Yahoo! Mail is a free email service offered by one of the most famous internet giants – Yahoo!

Excluding its Mail Business Email Plans, Yahoo! Mail is free and is the third-largest email service having over 280 million users as of December 2011.

The free version of Yahoo! Mail offers a huge 1 TB storage capacity, with an average email attachment limit of 25 MB. It supports POP3, IMAP4, SMTP, as well as mail forwarding depending on the country of the user.

Users in the US gained free access to their mail using POP3 as well as mail forwarding in 2013.

Yahoo! Business Email, on the other hand, is aimed at business users. With a $25 set-up fee and a $9.99 monthly fee, you get 10 accounts that can be managed by your administrator. The features include the following:

- Unlimited mail storage

- 10 email quota

- For up to 5 custom email addresses & a domain name, you will need to pay an additional $35

## 4. Zoho Mail

Zoho Mail is a rock-solid email service that provides lots of storage space along with POP and IMAP access. It targets professional users and offers online office suites and integration with instant messaging services.

Its free account allows up to a maximum of 10 users with 5 GB of space per user. Another 5 GB is reserved for document storage and is shared among the users. Now here's what's different with Zoho Mail: it requires that you use your OWN domain name – the email service will be provided by them but for your domain. Therefore, if your company does not want to get into the hassle of setting up and managing its own email server, Zoho Mail is definitely worth a look.

## 5. Inbox.com

Inbox.com has stood the test of time, evolving into a highly polished and reliable email service. POP3 can be used to access the Inbox.com mail account; however, as unfortunate as it may be, IMAP is not supported.

The web interface is pretty sleek in terms of performance, with rich text editing tools, free form labels, and threading. The solid spam and virus filtering features will ensure you get the least amount of spam and that your mails are free from viruses.

A 2 GB of online storage is also offered.

# Chapter 8: Email and Law

When it comes to email marketing, there are various laws in place, including the Data Protection Law. If these laws are breached, significant fines can be imposed upon the marketer.

The Irish data protection watchdog has, for instance, implemented strict laws for email marketing. Failing to adhere to these laws can lead to fines as high as €5,000 for each unsolicited email. While these laws are complex, the rules outlined below will help you better understand what you can legally do and what you should avoid.

## **DO:**

Whenever you are sending email to people who are not currently your customers, make sure that:

- These individuals have ASKED to receive such messages; AND
- They have been given an option to opt-out or unsubscribe from any further emails (and that too free of charge)

Whenever you are sending emails to people who are your customers, make sure that:

- The product or the service that you are marketing is similar to the one they have purchased from you; AND
- They were given an option to opt-out of receiving any marketing content when their details were being collected; AND
- They have been given an option to opt-out or unsubscribe from any further emails (and that too free of charge); AND
- A marketing email is sent to these customers every year with a clear option to unsubscribe – but they have chosen otherwise

Also, it is important to keep all the personal contact details of people on your marketing list completely secure, totally safeguarding their privacy.

All your employees who are responsible for carrying out marketing campaigns must be trained to do so in accordance with law, particularly data protection law.

**DO NOT:**

- Never ignore the rules outlined above
- You should never ignore any type of correspondence that you receive from a person who is complaining about receiving your marketing emails. There is no charge whatsoever for a person to complain to Ireland's data protection watchdog and an investigation would be carried out for the person
- Never pass a person's contact details or any other details without their consent to any third party
- Never buy marketing lists without confirming that the right consents have been collected from the people on the list

## Direct Marketing – Doing it legally

Direct marketing is when an individual is informed about a company's products and service. It is a legal activity as long as certain steps are taken to respect a person's privacy. Sending out unwanted marketing material is not good for a company and typically does not develop an interest in people receiving it.

Data protection laws put forth strict rules and obligations on the use of an individual's personal data. Due to the fact that marketing through electronic means is far more intrusive than the postal marketing methods, there are stricter rules to govern it.

## General Rule

The general rule is quite simple and straightforward. It states that anyone involved in direct marketing 'must have the consent of the individual (to whom they are sending the marketing material) to use their personal information for marketing purposes.

As a bare minimum, the individual must be given an opportunity to refuse to receive any marketing emails. Also, every email sent to them should have an option that allows them to opt-out free of charge.

Where it is not obvious, you would also need to clarify who you are and where you got the details of the individual.

# CAN-SPAM Act of 2003

On December 16th, 2003, President George W. Bush signed the CAN-SPAM Act of 2003 into the law. This act established the US's first ever national standard for commercial emails and empowers the FTC (Federal Trade Commission) to enforce it.

CAN-SPAM is an acronym that has been derived from this bill's full name. The full name is *Controlling the Assault of Non-Solicited Pornography And Marketing Act of 2003'.*

Critics often refer to this act as 'You-Can-Spam' because of the fact that it fails to cap many forms of spam, obstructing many state laws which would otherwise have empowered victims with the practical means of fighting spam.

For example, it does not demand email marketers to take permission before they send out marketing messages, hence encouraging bulk, unsolicited marketing emails. It also puts a restriction on states from implementing strong anti-spam protections, while prohibiting individuals who are receiving spam from being able to sue the spammers, except under the laws not applicable to email. As a result, the act remains to be largely unenforced even despite a letter sent to FTC by Senator Burns who stated, *"Enforcement is key regarding the CAN-SPAM legislation." (Wikipedia)*

Later on, several modifications were carried out to the original CAN-SPAM Act of 2003:

1. A definition of the term 'person' was added
2. The term 'sender' was modified
3. It was clarified that the sender can comply with CAN-SPAM act by including a private mailbox or PO box; AND
4. It was clarified that a recipient must not have to pay a fee in order to submit an opt-out request.

# Email Spam Legislation in Different Countries

The table below outlines the countries that restrict email spam.

| Country | Legislation | Section | Implemented |
|---|---|---|---|
| **Argentina** | Personal Data Protection Act (2000) | §27 | October 30, 2000 |
| **Australia** | Spam Act 2003 | Part 2 | 12 December 2003 |
| **Austria** | Austrian Telecommunications Act 1997 | §107 | |
| **Belgium** | Loi du 11 mars 2003 *("Law of March 11 2003")* | | 27 March 2003 |
| **Brazil** | None *(loosely; Movimento Brasileiro de Combate ao Spam)* | | |
| **Canada** | Personal Information Protection and Electronic Documents Act 2000 (PIPEDA) | | |
| **Canada** | Fighting Internet and Wireless Spam Act 2010 | | |
| **Canada** | Canada's Anti-Spam Legislation 2014 (CASL) | | |

| China | Regulations on Internet email Services | | 30 March 2006 |
|---|---|---|---|
| Cyprus | Regulation of Electronic Communications and Postal Services Law of 2004 | §6 | |
| Czech Republic | Act No. 480/2004 Coll., on Certain Information Society Services | §7 | |
| Denmark | Danish marketing practices act | §6 | |
| European Union | Directive on Privacy and Electronic Communications | Art.13 | 31 October 2003 |
| Finland | Act on Data Protection in Electronic Communications (516/2004) | | |
| France | Loi du 21 juin 2004 pour la confiance dans l'économie numérique *("Law of June 21 2004 for confidence in the digital economy")* | Art.22 | |
| Germany | Gesetz gegen Unlauteren Wettbewerb (UWG) | Art.7 | |

| | *("Act against Unfair Competition")* | | |
|---|---|---|---|
| **Hong Kong** | Unsolicited Electronic Messaging Ordinance | | 22 December 2007 |
| **Hungary** | Act CVIII of 2001 on Electronic Commerce | Art.14 | |
| **India** | None *(loosely; Information Technology Act, 2000 §67)* | | |
| **Indonesia** | Undang-undang Informasi dan Transaksi Elektronic (ITE) (*Internet Law*) | | |
| **Ireland** | European Communities (Electronic Communications Networks and Services) (Data Protection and Privacy) Regulations 2003 | Section 13 (1) (b) | 6 November 2003 |
| **Israel** | Communications Law (Telecommunications and Broadcasting), 1982 (Amendment 2008) | Art.30 | December 2008 |
| **Italy** | Data Protection Code (Legislative Decree no. 196/2003) | §130 | |

| | | | |
|---|---|---|---|
| **Japan** | The Law on Regulation of Transmission of Specified Electronic Mail | | April 2002 |
| **Malaysia** | Communications and Multimedia Act 1998 | | |
| **Malta** | Data Protection Act (CAP 440) | §10 | |
| **Mexico** | None | | |
| **Netherlands** | Dutch Telecommunications Act | Art.11.7 | |
| **New Zealand** | Unsolicited Electronic Messages Act 2007 | All | 5 September 2007 |
| **Pakistan** | Prevention of Electronic Crimes Ordinance 2007 | §14 | 31 December 2007 |
| **Russia** | None (*loosely: Russian Civil Code: Art.309*) | | |
| **Singapore** | Spam Control Act 2007 | | 15 June 2007 |
| **South Africa** | Electronic Communications and Transactions Act, 2002 | §45 | |
| **South Africa** | Consumer Protection Act, 2008 | §11 | October 2010 |

|  |  | | |
|---|---|---|---|
| **South Korea** | Act on Promotion of Information and Communication and Communications Network Utilization and Information Protection of 2001 | Art.50 | |
| **Spain** | Act 34/2002 of 11 July on Information Society Services and Electronic Commerce | | |
| **Sweden** | Marknadsföringslagen (1995:450) *"Swedish Marketing Act"* | §13b | |
| **Turkey** | Elektronik Ticaretin Düzenlenmesi Hakkında Kanun *"Act About Regulation of E-Commerce"* | | |
| **United Kingdom** | Privacy and Electronic Communications (EC Directive) Regulations 2003 | | |
| **United States** | Controlling the Assault of Non-Solicited Pornography and Marketing Act of 2003 (CAN-SPAM Act of 2003) | All | 16 December 2003 |

SOURCE: www.wikipedia.org

# Chapter 9: SPAM

Electronic spamming refers to the use of e-messaging systems to send out unwanted messages, particularly for the purpose of advertising. Even though the most common type of spam is e-mail spam, there are other mediums through which spam is sent out, such as through instant messengers, blogs, wikis, mobile phones, and others.

The focus of this chapter is to discuss in detail e-mail spam.

## What is Spam?

Email spam is commonly known as junk mail or as unsolicited bulk email (UBE), and it is the most common form of electronic spam almost identical messages that are sent to various recipients through email. Opening any links or attachments provided in the email can direct a user to websites that host malware, or to other phishing websites. Spam emails are also known to contain malware scripts as well as executable attachments.

Spam email is normally defined as email that is unsolicited and is sent in bulk quantities.

Since the 1990s, email spam has grown considerably as botnets and networks of infected computers have been increasingly utilized. It is estimated that these methods are used to send almost 80 percent of all spam.

## Types of Spam

1. Unsolicited Advertisements

Unsolicited emails, if found in bulk, can be quite a hindrance as they are likely to stack up one's entire spam folder. On a daily basis, thousands of such emails are sent for the sole purpose of advertisement. The products on offer usually include weight loss cures, knock-off merchandise, online programs for education, prescription drugs, and male enhancement pills. However, for the most part, these types of advertisements remain relatively low on the ladder of spam emails.

2.  Phishing Scams

Phishing scams are among the most difficult types of spam email to spot. Designed to appear as official emails sent from big enterprises or financial institutions, including the likes of PayPal and eBay, these emails are actually created to direct victims towards scam sites that seem equally official as compared to the originals, but in reality, are not. Due to this, people are tricked into volunteering their login credentials, which are afterwards used by the scammers for compromising the customers' actual accounts.

3.  Nigerian 419 Scams

This is one category of spam messages which each individual is certain to receive at one point or another. The Nigerian 419 Scams appear as seemingly amazing offers out of nowhere from a stranger who lives in a land far away. The email comes from a person who is claiming to be an agent, an employer, or a lottery service. The offer is always the same, a considerable amount of money, with a small percentage of it in return for the time, shipping, insurance, or any other apparently legitimate reason this alleged individual provides. The individual forwards a fake check upon agreement, asking for the small amount of cash to be wired back. Those who fall victims never receive the money from the check, and are deducted the small fee from their accounts, which is typically a few hundred dollars.

4.  Email Spoofing

This is one technique which is used for making other spam email tactics appear more believable. Spammers tend to send a message which appears to be originating from a varying email address. The email spoofing technique seems to appear as if a scam email has actually been received from a trusted organization, company, or source. Due to this, the victim trusts the message and likely falls prey to the scam, whatever it may be.

### 5. Trojan Horse Email

This is among the ancient techniques for email spam. The Trojan horse emails, or email worms, are scam messages which infect the computer of the victim. Moreover, upon infecting, these emails also send themselves out to all the emails in the individual's contact list. The most well-known email worm dates back to the year 2000, which was the "ILOVEYOU" bug. This seemingly came from an actual person (who would have fallen prey beforehand) and was titled "I love you." Upon downloading the script attached in it, one's local computer would be damaged by the Trojan horse virus.

### 6. Commercial Advertisements

Most reliable websites ask their audiences to sign up for their newsletters, advertisements, or other types of messages. Others, however, don't ask for any approval and automatically add their customers to their mailing lists just to get them on board for some type of offer or advertisement. This results in unsolicited bulk messages, most often ending up in spam folders.

### 7. Anti-Virus Spam

It is understandable that users with a lack of knowledge regarding the technical aspects of their computers are likely to fall very easily for any spam email. When this spam email claims to be an "anti-virus" which can "remove the virus their computer is infected with", the reasons become all the more strong for these individuals to end up as victims. Upon downloading the given security software, the victim's computer becomes infected with a virus. In most cases, the downloaded software would even go as far as demanding cash for magically cleaning up the virus it just infected the individual's computer with.

### 8. Chain Letters

Chain letters are the most obvious types of scam emails. These are sent by anyone one would already know and be familiar with, consisting of funny photos, recycles jokes, ancient prophecies, or sensational claims about the President. Chain letters can be easily spotted by the hundreds of forwards they will come attached with at the top of the mail.

### 9. Political or Terrorist Spam

Political or terrorist scam is intended to scare the victim, all the while stealing their personal information. The scammers who send these out claim to be a politician or any other well-renowned government agency, such as the FBI, stating that the individual who receives the mail is in danger. For getting rid of the threat, the scammers want personal information or even cash in exchange, when in truth, there is no such thing that qualifies as a threat to begin with.

### 10. Porn Spam

Pornography has been a huge business all around the world for a considerable amount of years now. The fact that it is used by a majority of the population across the globe adds to its likelihood of being utilized as a primary source for malicious content. Porn spammers purchase email addresses (or often harvest them) from people, using them to send out provocative advertisements consisting of inappropriate content. The victims are more often than not directed to adult sites as a result.

# Spam Filters

Email filtering is a process through which emails are organized based on a certain criteria. In most cases, spam filters perform an automated filtering of all incoming email. However, the term 'email filtering' also applies to the anti-spam techniques through human intervention.

Email filtering tools take in email, processing it and either:

- Passing the email for delivery to a user's mailbox, or

- Redirecting the mail for delivery elsewhere, or

- Deleting the message

Some of the email filters in use are also capable of editing messages when processing them.

## Uses of Mail Filters

Mail filters are normally used to organize the incoming mail and to remove spam or attached malware. A less frequent use of mail filters is to scan outgoing email at certain organizations to make sure that the employees are in compliance with any applicable laws of the organization with regards to the use of email.

Users may even use a mail filter to prioritize your messages, while sorting them into specific folders depending on its subject or numerous other criteria.

## Methods & Techniques

Mail filters can be installed manually by a user as an application or as an inherent part of their email client. Email clients, such as Outlook and Thunderbird allow users to make their own, customized mail filters that can handle the incoming email based on the specified criteria.

The majority of modern email clients now come with automatic spam filtering functions. Today, almost all ISPs install mail filters within their mail transfer agents to provide a spam filtering service to all their clients.

Because of the ever-increasing threat of phishing and other fraudulent websites, many ISPs now filter URLs from emails to prevent users from unknowingly opening potentially harmful websites.

## Inbound & Outbound Filtering

Mail filters can operate for both inbound as well as outbound email traffic. The former involves scanning of email as it arrives at the filtering system. The latter, on the other hand, employs the reverse process by scanning the email, which is being sent from the local network to other recipients on the internet.

One common technique of outbound email filtering that is often implemented by ISPs is known as *transparent SMTP proxying*, where the traffic is intercepted and then filtered through a transparent proxy inside the network.

Outbound mail filtering can also happen at an email server. A number of organizations utilize data leak prevention technologies within their mail servers to avoid leakage of crucial information through email.

# How to Not Get Blacklisted

With the frequency of spam that circulates the Internet in huge amounts, it is becoming more and more crucial for system administrators to properly understand the reasons that can cause their IP address to end up on any blacklist. Spammers use all types of methods and tricks to send as many spam emails they can without disclosing their real identities. They accomplish this through a number of techniques including social engineering, botnets, malware, by forging message headers, and through exploitation of weaknesses in email servers and network infrastructures.

For a spammer, this is just a game of numbers. It costs them almost nothing to send thousands and thousands of spam email, and if even a tiny number of individuals purchase their product that is being advertised in the spam email, or click on a link, the spammer can gain considerable profits. With that said, If your mail infrastructure has not yet been properly and thoroughly secured, then it is at great risk of being infected, thereby becoming a part of the spammers' botnet.

Regardless of whether your server is infected with malware or not, if the mail server security settings or firewall has not been configured properly, the IP address of your server can end up on a blacklist.

To prevent your server from being blacklisted, ponder over the following factors:

- **Necessitate strong passwords** – Spammers often perform dictionary attacks on mail servers all the time. A dictionary attack makes use of a large list of commonly used words that are used as passwords in an attempt to guess the correct password and gain control of an account. To fight this, the users who have accounts on your mail server must always employ strong passwords. Easy to guess passwords such as "1password" must be avoided at all costs. Users should be encouraged to create passwords containing both lowercase and uppercase, numbers, and other characters. In MDaemon, for instance, you can make it compulsory to have strong passwords by accessing Accounts menu, Account Options and Passwords menu.

- **Use SMTP Authentication** – It is recommended that all the users utilize SMTP authentication method. In MDaemon, this can be done so by going to Security, Security Settings, Sending Authentication, and SMTP Authentication. Check the box that says *'Authentication is always required when mail is from local accounts."* Ensure that '…unless message is to a local account' has been unchecked.

- **Do Not Permit Relaying** – Relaying occurs whenever a mail that is neither to, and nor from, any local account, and it is sent through your email server. It is extremely common for spammers to utilize open relays. Hence, make sure that your mail server does not relay any mail. This can be done in MDaemon by going to Security, Security Settings, Relay Control, and checking the boxes below:
  - o **Do not allow message relaying**
  - o **SMTP MAIL address must exist if it uses a local domain**
  - o **SMTP RCPT address must exist if it uses a local domain**

**It is recommended that you do not check the exclusion boxes displayed on the screen. (SOURCE: www.mdaemon.com)**

- **Ensure Your PTR Record is Valid** – Your PTR record should match your outbound public IP to the name of your mail server, or to a fully qualified domain, or an FQDN. An ISP can create such a record for you. A PTR record enables servers to carry out a reverse DNS lookup on the IP address that is connecting in order to verify that the name of the server is linked with the IP address from where a connection has been initiated.

- **Set an SPF Record** – Sender Policy Framework, or SPF is a type of anti-spoofing technique, which determines whether an incoming mail from a particular domain has been sent from a host that had authorized the mail to be sent from that domain. In other words, it is an opposite of the MX record – whose method is to specify hosts that have been authorized to collect mail for a specific domain.
- **Configure IP Shielding** – IP shielding is a unique security feature that enables you to specify the addresses or range of IP addresses that are authorized to send mail for a certain domain. Make sure you configure the IP shield on your server to only accept the incoming mail from local domain if it is received from an allowed IP address.
- **Enable Secure Sockets Layer (SSL)** – SSL is an encryption method used for encrypting a connection between the server and a mail client. Ensure that all those mail clients that are connecting to the server use SSL ports (MSA 587, SMTP 465, POP 995, and IMAP 993).
- **Use Account Hijack Detection Feature** – Some server software, such as MDaemon, have an account hijack detection feature that can put a cap on the number of emails a certain account can send in a specific amount of time. This feature typically applies only to authenticated sessions, and it is utilized to prevent an account that has been compromised from being used to send huge amounts of email spam.
- **Use Dynamic Screening** – Dynamic Screening, just like account hijack detection feature, may be used to block connections from certain IP addresses depending on a certain criteria, such as the behaviour of activity from a particular IP or IPs.
- **Sign with DKIM** – DomainKeys Identified Mail or DKIM is designed to help and protect users against theft of email

identity and from tampering of email contents. It achieves this by positively identifying a signer's identity with the encrypted 'hash' of email content. With DKIM, two keys are created: a private one and a public key. The latter is published to the DNS records of the signing domain, while outbound messages get signed by the private key. The receiving server then reads the key from a DKIM-Signature in the email's header, comparing it with a public key found inside DNS records of the sending domain.

- **Block Port 25** – Set your firewall to only permit outbound connections on the port 25 from spam filter or mail server. No other systems on the network should be permitted to send data outbound on port 25.

## How to Get On a Certified Whitelist

'Certified email' is a common technique used for email whitelisting purposes. It is a process by which the ISP enables an individual to bypass their spam filters when an email is being sent to any of its service subscribers. This service typically comes at a certain cost for the subscriber. A certified whitelist provides the subscriber the peace of mind that important emails (that have been whitelisted) will not be considered spam. Also, it will ensure that no links or images are stripped from the emails by the automatic spam filters.

The primary aim of the certified email is to enable organizations to reach out to their customers in a reliable manner, while allowing the recipients the assurance that their 'certified' message is indeed legitimate and not an attempt at phishing.

An email that is certified by a neutral 3rd party is also referred to as certified mail.

## Certified Email

CertifiedEmail is a certified email service by Goodmail Systems. It is one of the most common yet controversial service that has appeared in headlines since the announcement of AOL and Yahoo! with regards to its implementation within their mail servers.

According to AOL, email from those senders who are in the qualified list and have paid 10 cents per email will bypass spam filters and will be delivered directly to the mailbox. AOL also announced that it will pay the charges for all non-profit organizations. Such messages will be marked as those coming from a source that is trustable.

The senders who wish to have their domain whitelisted will have to go through an accreditation system with the developer of CertifiedEmail, Goodmail. Additionally, their emails may only be sent to those individuals who already have an existing business relation with the email's sender. If a particular sender emails an individual who has not agreed to receive the email, AOL may completely block the sender.

AOL states that the free email service will continue to work as it did previously, and users will get all the emails from those senders whom they have whitelisted. No charges will be incurred on the users for either sending or receiving email. However, those who do not 'prepay' for the whitelisting service CertifiedEmail – all of their email will be subjected to scrutiny by spam filters.

It is important to learn about the controversial issues surrounding Goodmail's CertifiedEmail service. MoveOn, a public policy advocacy group, protested AOL's use of CertifiedEmail. It labeled the latter program as 'email tax', claiming that AOL is offering spammers a direct entry to the mailboxes of users, while making people shift to paid services by filtering out email that is legitimate (but not paid for).

The following ISPs in the United States have adopted CertifiedEmail:

- AOL
- AT&T
- Comcast
- Cox
- Road Runner
- Verizon
- Yahoo

## What You Should Know About Email Whitelists

All email marketers are aware of, and dread, blacklists. Not everyone fully understands the criteria that is used to blacklist people; however, what is clear is the fact that if your domain is on a blacklist of a certain mail server, your messages will not be delivered to the intended recipients using that specific mail service.

Does the opposite of blacklists do any good to the sender? Do whitelists and certifications really give you a status that is special and ensures that your mail is delivered? Let's find out.

### Good Sender Record

Before we can even begin to answer the above questions, we have to be clear on the differences between certification services and whitelists.

## Certification

Certification services are offered by 3rd party reputation service providers (RSPs) that warrants the mail from IP addresses and domains of senders that have been qualified. When the receiving ISP recognizes such mail, a certification service identifies senders and exempts the email from some or all of the antispam filters/rules.

In certain cases, extra privileges may also be offered, including inbox placement, trust tokens, and image/link rendering.

RSPs normally charge a certain transactional or subscription fee to all the qualified senders.

Some have likened the concept of email authentication to the licence registration of a vehicle. It implies that a person owns the vehicle and is authorized to drive it on road; however, it does not state anything about the driving record of the individual. A test has to be passed to get the licence and to maintain it, a good driving record has to be ensured. However, the license does not guarantee that the person will not be involved in any kind of an accident.

Certification is very similar to an insurance policy for safe drivers. Depending on who is issuing the insurance policy, the person may get additional privileges; however, no matter how much 'premium' is paid, the certification states might be revoked if a good driving record is not maintained.

The availability as well as the benefits of certification services and whitelists are different according to RSP and ISP. Some ISPs, such as Yahoo and AOL, have their own whitelists while utilizing certification services to offer more privileges to qualified senders. RoadRunner and Windows Live make use of a certification service in place of a whitelist. Other ISPs usually maintain a whitelist or utilize an antispam filter that takes into account the certification along with various other factors found in its algorithm.

## Qualifying for Certification and Whitelisting

The requirements for qualifying for whitelisting and certification vary from one service provider to another. However, some of the common points are mentioned below:

- **Customer Satisfaction** – Low rates of customer complaints is essential for all certification services and whitelists. Therefore, all those factors that have a strong effect on complaints is extremely crucial. For many services, opt-in consent is a necessity, and they go a step further to examine data capture as well as the permission practices of the sender. In certain cases, a sender may be required to substantiate permission in case of customer complaints. The privacy policy and the handling of complaints and requests for unsubscribing may also be looked into.
- **List Management** – A number of ISPs have a requirement for efficient bounce management. A low level of 'unknown users' is a key point.
- **Regulatory Compliance** – Strict compliance to the 'Can Spam Act' is a must. There should be no attempts to hide, misrepresent, or forge the sender or to deceive the receiver with regards to the intention of the email. All the emails must have a valid, non-electric contact information of the sender. Not only that, but a simple unsubscribe mechanism should be present allowing users to easily unsubscribe. The sender should honour all unsubscribe requests within a reasonable time frame.
- **Email Authentication** – Many services require that the sender authenticate its email using the protocols used by the receiving mail server, such as DKIM, SPF, or Sender ID.
- **Sending Infrastructure** – It is a core requirement of all ISPs that the sender's email infrastructure is operated in a responsible way, and is well maintained. This means that email servers connecting to an ISP must have their valid

reverse DNS entries and must be secured to prevent any unauthorized use. Static IP addresses is also a requirement in many cases.

## Getting Started

Let us assume that you have met all the requirements as stated above. What should be the next step? Should you go for ISP whitelisting? Or should you invest in RSP certification?

As far as the first question goes: yes, you should go for ISP whitelisting as it is free. You have nothing at stake here. If your domain is on an ISP whitelist, this will prevent your emails from being filtered as spam, even if your practices or complaint rate deteriorate.

However, do not forget the fact that whitelisting will classify you as being a legitimate sender in the eyes of the ISP, thereby opening an crucial channel of communication even if an "accident" was to take place. In addition to that, the qualification process for getting whitelisted will help you gain a better understanding of the various requirements of numerous ISPs; hence, you will be in a better position to revamp your own standards.

When it comes to the question about RSP certification, things get a little complicated. To begin with, you will need to thoroughly examine not only your delivery rate, but also a number of other key metrics such as opens and clicks. This should be done once you have taken advantage of whitelists, to further improve performance.

Due to the fact that the coverage and the benefits offered by numerous ISPs vary, you should review not only the performance, but the distribution of your whitelist by an ISP in order to compare what other RSPs offer. If you have a solid reputation and a high delivery rate, you may discover that you will not benefit too much by getting certification from an RSP in terms of deliverability. In fact, the areas that may see improvements are the image/link rendering and better placement, leading to higher clicks and greater conversion rates.

Some of the services also provide enhanced solutions for B2C mailers instead of B2B or vice versa. The final and crucial step is the ROI testing. You have to make sure that the amount of money you are putting in to ensure your email is delivered is really worth it and will give you the highest possible returns. In a nutshell, you should only go for an RSP certification if you find that it will have a positive impact on your bottom-line results.

Just like email authentication, certification services and whitelists have become an integral part of the new email ecosystem. These services assist by conveying the senders' reputation. Once the identity of the sender has been established through the authentication process, the ISPs can do a much better job of correctly sorting out good email from the spam, thereby preserving the medium for just legitimate communication & commerce.

## Do You Really Need Certification Services?

Whitelists have always been a part of discussion when the topic of email and spam comes up. Just to quickly recap, whitelists have a simple and easy to understand concept:

The sender, that is you, gets his or her IP addresses in an ISP's whitelist. This allows the email sent by you to be delivered into the mailbox without having to go through spam filters, which may sometimes result in false positives.

However, over the past few years we have seen the concept of whitelisting being merged – along with other best practices – into a concoction of services called 'certification services'. The concept of the latter services is almost identical to and simple enough like whitelisting: the company that offers the certification services has contracts with a number of ISPs. Therefore, any IP address that ends up in the whitelist of this company (we will call it RSP), will be seen by all the ISPs that have agreements with the service provider. This will not completely exempt the senders' messages from being scrutinized by the spam filters; however, the antispam filters will definitely be more lenient while providing many advantages such as the messages being image/link enabled.

In order to certify and end up on the list of an RSP, the sender has to fulfil and meet certain standards – these standards are commonly referred to as 'best practices'. Some of these 'best practices' include:

- Double opt-in
- Unsubscribes exclusion
- List hygiene
- Transparency with recipient
- Some other factors

The two most important questions about certification services that will help clear up a lot:

- **Are these certification services nothing but expensive whitelists?**

This is not true. You cannot buy your way into certification services whitelist. You have to 'try' and get your IP addresses and domain certified. For this, you will have to be a 'great sender' that adheres to best practices and respects the privacy of recipients. So if you have what it takes to get certified, it is generally a good idea to go for it particularly if your marketing strategy comprises of email marketing.

All in all, getting certified can have a great effect on your business's bottom line, and the cost can be well worth it in most cases.

- **Do you really need them?**

Whether or not you need certification services depends on a number of things. It is best to consider your domain database distribution, volume figures, and of course, your budget. If your organization relies heavily on email marketing, then it would make sense to take every possible measure to ensure that you are on whitelists, and not black ones!

## AOL Whitelist Information

AOL provides information regarding its whitelist for mail server administrators. The AOL whitelist has been designed to help better collaboration between AOIL and individuals or organizations that are involved in sending solicited emails. It also aims at protecting AOL users from unwanted messages. Hence, to cater to this, AOL offers two types of whitelists:

- **AOL Standard Whitelist**
- **AOL Enhanced Whitelist**

### AOL Standard Whitelist

The standard whitelists allows messages to be delivered to the mailbox of users by bypassing some (not all) of AOL's antispam filters. The protection provided by this whitelist may be revoked at any moment without any warning if AOL has a reason to believe that a certain IP no longer satisfies the qualification criteria. A whitelist application has to be filled by those who wish to be considered for an entry into the list.

### AOL Enhanced Whitelist

The Enhanced Whitelist (EWL) allows emails with links and images to be delivered to mailboxes without stripping them. The EWL is only available for those who are already in the standard whitelist. These IPs are evaluated for entry into the enhanced whitelist on a daily basis. EWL is only granted based on the sender's IP's reputation.

Every IP is evaluated on an individual basis, and qualifying IPs do not extend EWL status to their complete domain.

# Sender Score Certified

Return Path, a company based in New York, offers Sender Score Certified – an email accreditation service. Once a sender is accredited, their IP addresses are added to Sender Score Certified whitelist. All ISPs, email service provides, and antispam technologies that use Sender Score Certified whitelist will offer preferential treatment to the emails sent from the whitelisted IP.

Accreditation is granted only after the sender reaches certain preset standards. An audit is carried out to determine the sender's email marketing reputation as well as practices. All IP addresses that have been accredited are continuously monitored; a failure to maintain the required standards can result in the whitelisting being revoked or suspended.

An annual fee applies to keep the Sender Score Certified account in operation.

**Benefits of Sender Score Certified**

- Improved deliverability to recipients
- Added benefits for Windows Live Hotmail, including automatic rendering of links and images, inclusion of unsubscribe button within the web interface
- No hourly throttling limits
- Lenient daily throttling limits

Regarding the Sender Score Certified, Ken Takahashi, Return Path's VP, says:

*"Sender Score Certified is the largest, most comprehensive and widely used whitelist covering over 1.2 billion inboxes of the world's top ISPs. This program is designed for commercial senders who are willing and able to maintain the highest ethical, professional, and technical standards when sending email. We cover Windows Live Hotmail, Road Runner, GoDaddy, Time Warner Cable, USA.net and many more. We also cover spam filters from Barracuda, Cloudmark, IronPort and Spam Assassin."*

*"Also, because of our reputation data network, we know that email on Sender Score Certified meets higher standards than our competitors meaning that is has lower complaint rates and fewer spam trap hits. Clients see up to 17% lift on delivery when they come onto Sender Score Certified."*

**Source:** http://www.email-marketing-reports.com/deliverability/certification/senderscorecertified.htm

## Certified Senders Alliance

Senders of bulk emails who are involved in sending advertising material with the permission of the recipient often complain about the reduced deliverability, stating that solicited messages such as newsletters do not reach the inbox of the intended recipients. Currently, such bulk mailers deal with this situation by negotiating with ISPs in order to be added to whitelists.

This, however, has numerous disadvantages. For starters, a bulk mailer has to negotiate with many ISPs. On the other hand, the ISPs themselves have to scrutinize many bulk mailers to ensure that they meet the standards and practices outlined by the ISPs. These standards also vary from one ISP to another – making it a complex task for bulk mailers to cater for all requirements of individual ISPs. To add to the problem, changing contact people further make the process more complex.

### The Proposed Solution

Such situation demands setting up of a centrally managed whitelist consisting of bulk mailers. The whitelist should possess standardized procedures to simplify the approval process. A centralised whitelist not only effectively provides an efficient Spam fence, but it also acts as an interface between bulk email marketers and numerous ISPs – making the task easier for both the parties. The purpose of this centralised list is to enhance the quality of the medium called 'email', while ensuring that wanted emails sent by legitimate senders (both ways) are delivered to the intended recipients.

Participation in the CSA (Certified Senders Alliance) means that the spam filters of the server are typically not activated; hence, providing the recipient with greater control of which bulk emails they wish to receive.

This is precisely why eco e.V. and the German Dialogue Marketing Association e.V. cooperated and worked together so the whitelist project could benefit from the backing of Internet industry along with support from direct marketers.

This service is completely free of charge for technology partners and ISPs, whereas members of DDV, BCDW, and eco are entitled to a 20 percent discount on certification price.

## Benefits of CSA Certification for Marketers

Bulk mailers can enjoy a number of benefits by gaining CSA certification, including the following:

- Efficient email delivery
  - CSA acts as the central contact point for all whitelisting purposes;
  - 100% server-sided delivery;
  - No need for spam-filtering or spam tags from the provider.
- Highest quality criteria
  - Legal conformity for email marketing is ensured by certification;
  - Transparent and standardized processes;
- One-stop solution for complaint management
  - CSA maintains a central hotline for complaints;
  - In case of any issues, CSA acts as the interface to ISPs;
  - Senders receive notifications of any violations immediately.

# Chapter 10: Email Security

Internet and e-mail have taken communications to a whole new level. Most of the Internet users heavily rely on e-mail for their communication needs. However, while email has made it easy for people to keep in touch – for both personal and professional purposes, it has also become the most common way of spreading viruses, worms, and Trojans.

Unfortunately, malware is not the only factor that haunts all those who use email. Email privacy has always been a major email security consideration. Email can be intercepted and screened, for instance, when a business wants to monitor its employees. If this is not a case, there is no shortage of cyber criminals prying on sensitive information contained in email messages.

All of these privacy issues involve countless ethical, legal, and moral issues. This is precisely why steps need to be taken to fully grasp the topic of email security, and understand the available solutions to the existing problems.

## Email System Vulnerabilities

### 1. In Storage

Separate messages that are saved on hard disks as text files are typically neither compressed nor encrypted. This means that the text file can be accessed and read by anyone can access the computer where the messages are saved. These messages can be protected by the security of the operating system's file system; nevertheless, they are susceptible to viewing.

In LAN-based systems such as Groupwise, Exchange, and Notes, such messages are saved in databases. The main benefit of storage databases is their high level of efficiency and their ability to offer an added layer of security.

Even with this, the system is still not protected from, say, a rogue network administrator who can access all of the software and possesses the security clearances required to access the database.

Overall, servers have enhanced file system security as compared to workstations. Thus, only those people with appropriate clearance can view files in any specific directory. On the other hand, servers are also protected by heavy firewalls, backup systems as well as physical protection from theft and fire.

## 2. During the Transport

When messages are being transported from one computer to another, there is a huge chance of these messages being intercepted and viewed by unauthorized persons within the same network segment. This is referred to as 'packet sniffing'. This threat is not posed just by external factors, but also from within the organization. For instance, an individual within the organization's LAN may run a sniffer application and monitor the email traffic across the network. The problem is further intensified with the free availability of sniffer programs; while they do require good computer knowledge and skills, the threat is still serious.

Packet sniffing is not the only way by which the emails being transported can be viewed. Once the message is transmitted from one system to another, it is copied to a certain queue on the hard disk. This can be copied by someone who has access to the system, and kept for future review.

# Email Security Techniques and Guidelines

Emails may be intercepted quite easily, therefore requiring that certain techniques and guidelines be employed to ensure their safety. Even though this book is aimed at system administrators and network specialists who have advanced knowledge of security systems, it would be a good idea to quickly review a few 'common-sense' defence techniques – you may wish to pass these on to all the users on your system.

- Regularly change email passwords and share them with no one. Server administrators and ISPs never ask their users for their passwords. If you receive any such emails, do not respond to them.

- Do not open any executable attachments without scanning them with anti-virus software. Many viruses and worms tend to disguise themselves in extensions such as .scr, .exe, .vbs, and others.

- Scan all the incoming emails. This can be done by a gateway anti-virus (installed as a part of your email architecture).

- Exercise caution when opening email messages from an unknown origin. In corporations, if the email appears to be highly suspicious, it should be forwarded to the system administrator immediately for further analysis.

- Never respond to spam. This may not appear to be a security issue, but this practice can limit the exposure you have to unwanted email messages. Responding to spam to have the email address removed from the list actually validates that an account is valid, putting it in the spammer's 'A-list'.

- The latest security patches should be applied to system(s) to seal any vulnerability. As a system administrator, you should apply these to the mail server to protect it from being exploited in any manner whatsoever.

## Connection Protection

Computers have to agree on predefined set of protocols or instructions before connecting to each other and for transmitting the messages. Protocols such as POP3, SMTP, and IMAP are 'clear text' methods of transmission. Anyone using a sniffer program can view the passing traffic, and read the messages. Use of encryption methods can assure safety of a message even if it is intercepted – encryption prevents the unauthorized person from viewing the actual content of the email in it intended form. Encryption will be discussed in detail later in this chapter.

## Security Features of Microsoft Outlook

Microsoft Outlook is a solution that allows organizing and managing of digital communication tools, including email, instant messaging, and newsgroups together with calendars, contacts, notes, and lists. Outlook fully controls the email messages, contacts, and appointments, thereby allowing the user to effectively manage their time as well as tasks.

With that said, here is a brief overview of security features of Microsoft Outlook:

1. **Encryption**

A user can mark any message for encryption when the message is being composed. Outlook utilizes the public key of the receiver to encrypt the message. The reciever's client then uses the private key to correctly decrypt the message and show it to the authorized user. This ensures confidentiality as a person without the receiver's key cannot decrypt the message and view its content.

### 2. Digital Signature

A user can digitally sign an email message in Outlook. Outlook enables encryption of message with sender's private key, and then appends the public key into the message so that the receiver does not have to look for it in the directory. To prevent problems such as impersonation, the key is signed by a certificate authority, such as VeriSign. Digital Signing will be discussed in greater detail later in the chapter.

### 3. Email Attachment Screening

Microsoft Outlook is known to automatically block emails that contain executable files as attachments. These executable extensions include .scr, .exe, .vbs, and others – and these are able to carry malware. It warns the user whenever they attempt to open such files or when they read mail that contains scripts. This method, however, has drawn criticism from a lot of users who believe intelligent users should be empowered to take security decisions.

### 4. Spam Protection

If an outside program attempts to utilize the 'Send' feature without the consent of a user, a flag would be raised and the user would be prompted to confirm whether he or she has allowed the outgoing mail. In addition, to manage dictionary attacks, Outlook features a junk-filter that can delete incoming spam and prevent it from entering the Inbox folder.

# Encryption

Email messages can be encrypted to protect them from being viewed by anyone but the authorized individuals. It is a method used to counter privacy issues that we discussed previously in this chapter.

Email encryption methods may rely on the public-key cryptography – a technique where users publish a public key that can be used by others to encrypt messages for them. On the receiving end, these users possess a private key that is needed for decrypting messages.

## Understanding Encryption Protocols

The protocols used for email encryption include the following:

1. Identity-based encryption

2. Bitmessage

3. Mail sessions encryption

4. OpenPGP

5. TLS

6. S/MIME

# 1. Identity-based Encryption (IBE)

IBE is an offshoot of ID based cryptography in which a public-key is based on a unique piece of information related to the user, such as their email address. This uses the text value of a name or a domain name as the key or as a physical IP address. The concept of ID based encryption was proposed in the 1984 by Adi Sharma. He, however, could only give instantiation of identity-based signatures. Therefore, IBE remained a problem for several years.

Nevertheless, in 2001, the Cock's encryption scheme and pairing based Boneh-Franklin scheme solved the IBE problem.

IBE offers a unique approach towards management of encryption keys. It utilizes an arbitrary string as the public key, allowing protection of data without requiring any certificates.

# 2. Bitmessage

Bitmessage is an encrypted, decentralized, peer-to-peer communication protocol that can be used by any person to send an encrypted message to another individual. Bitmessage is capable of encrypting each of the users' mailboxes through public-key cryptography, replicating it within its own P2P network and mixing it with other mailboxes to hide a user's identity, thwart eavesdropping, and enable the network to run in a decentralized way.

The Bitmessage protocol, through authentication, avoids sender-spoofing and conceals the metadata from various wiretapping systems.

# 3. Mail sessions encryption

The STARTTLS SMTP is a type of TLS layer superimposed upon an SMTP connection. Even though it protects the traffic from packet sniffers, it is not a way of encrypting emails because of the fact that content of email can be viewed and altered by any intermediate email relays along the way. Putting it in simpler words, the encryption tends to apply between each of the SMTP relays – not between the sender of the email and its recipient.

If both relays have support for STARTTLS, it can be utilized regardless of whether the email content has been encrypted by any other method of encryption.

STARTTLS is an extension of POP3 and IMAP4 as well. For more information, refer to RFC 4616.

## 4. **OpenPGP**

PGP or Pretty Good Privacy is a type of data encryption program that offers authentication and cryptography privacy for data communication. PGP is regularly utilized for encrypting, decrypting, and signing emails, texts, directories, and files.

It is heavily used for securing email communications ever since it was developed by Phil Zimmermann when he was working for PKWARE Inc.

PGP follows the OpenPGP standard (check RFC 4880) for encryption and decryption of data.

PGP encryption jointly uses data compression, hashing, symmetric-key cryptography, and public key cryptography. Each step also utilizes one of many supported algorithms. A lot of email clients have support for OpenPGP protocol.

It is important to distinguish between PGP and OpenPGP in the sense that the latter was developed to overcome patent issues.

## 5. **TLS**

TLS  or Transport Layer Security is a cryptographic protocol developed to offer secure communication over computer networks. It utilizes X.509 certificates and, in turn, asymmetric cryptography for counterparty authentication as well as for exchanging a symmetric key.

The protocol enables client-server applications to communicate over a network in a way that eliminates the possibility of tampering or eavesdropping. Due to the fact that protocols can function with or without the TLS, it is crucial for a client to specify to a server whether a TLS connection is required. There are two methods of accomplishing this.

Firstly, a different port number can be used to set-up a TLS connection. For instance, the port 443 is used for HTTPS.

Another way to do this is to have the client request a server to switch over to a TLS connection through use of a protocol-specific mechanism, such as using STARTTLS for news and mail protocols.

Once both the client and the server have agreed to utilize TLS for the connection, they establish a connection using handshake procedure.

During the handshake procedure, the client and the server agree upon the use of numerous parameters for establishing the connection's security:

1. The handshake starts whenever a client contacts a server (TLS-enabled) to establish a secure connection. The client presents a list of the supported hash functions and ciphers.
2. From this list, the server chooses a hash function and a cipher, which it has support for, and informs the client of the selection.
3. The server then responds by sending back its identification as a digital certificate. Included in this certificate is the name of the server, the name of the trusted certificate authority, and the public encryption key of the server.
4. The client can contact the server to confirm the validity of the certificate it sent before proceeding with the procedure.

5. To generate the required session keys for use with the secure connection, the client tends to encrypt a random number using the public key of the server, sending this to the server. The server can decrypt this using its own private key.
6. From this random number, both the server and the client can generate the material for the purposes encryption and decryption.

This brings the handshake procedure to an end and a secure connection is established, one that can be encrypted or decrypted using the key until the connection is closed.

If during the handshake procedure, any of the steps fail, the entire handshake process fails and the connection will not be established.

## 6. **S/MIME**

S/MIME stands for Secure/Multipurpose Internet Mail Extensions, and it is a standard for signing of MIME data and for public key encryption. It was originally created by RSA, and used the IETF MIME specification.

S/MIME offers cryptographic security services listed below for e-messaging applications:

- Authentication
- Message Integrity
- Non-Repudiation of Origin (through use of digital signatures)
- Privacy and Data security (through encryption)

S/MIME defines the MIME type for data encrypting, whereas the entire MIME entity that needs to be enveloped is duly encrypted and then packed into the object that is eventually inserted into application/pkcs7-mime MIME entity.

### **S/MIME Certification**

One has to acquire and install a certificate or key from a public CA or an in-house CA before S/MIME can be used for any type of application. The normal practice is to utilize distinct private keys (along with relevant certificates) for the signature and for encryption purposes. This allows escrow of an encryption key without compromising to the signature key's non-repudiation property.

# Digital Signing

A digital signature – or the process of digital signing is a scheme that is used to demonstrate the validity and authenticity of a digital document or a message. A valid signature provides the recipient a reason to suppose that the message was indeed developed by a recognized sender, and that the sender does not deny having sent the document/message. A digital signature also guarantees that the message was not altered while being transmitted.

Typically, digital signatures are utilized for financial transactions, software distribution, and numerous other instances where it is crucial to detect tampering and forgery.

**Digital Signature Scheme**

A digital signature scheme includes three (3) algorithms:

- **Key Generation Algorithm** – this algorithm chooses the private key at random from a number of possibilities. It then outputs a private key along with its relevant public key.

- **Signing Algorithm** – once this algorithm is provided with a private key and a message, it develops a signature.

- **Signature Verifying Algorithm** – once this algorithm is provided a public key, a message, and a signature, it accepts or rejects a particular message's authenticity claim.

Two crucial properties are needed. First of all, the authenticity of a particular signature that is generated from a fixed private key and a fixed message may be verified through a matching public key.

Secondly, it must be infeasible computationally to generate a applicable signature for a certain party without being aware of the relevant party's private key.

A digital signature is therefore an authentication mechanism, which enables a message's creator to attach a particular code that can act as the signature. It is formulated through a hash of the message and by encrypting a message with the owner's private key.

## Setting and Using Encryption for Email

The majority of full-featured email clients such as Microsoft Outlook, Mozilla Thunderbird, and Apple Mail have native support for the S/MIME secured email – which relies on digital signing and encryption of messages through certificates.

Some other methods for encryption include the PGP, and the GNU Privacy Guard (GnuPG). There are numerous software available for free and for fee that provide encryption for email messages. These include PGP Desktop Email or Gpg4win – both of which have support for the OpenPGP.

Even though PGP can provide protection to the messages, it can be quite complex to correctly use. Back in 1999, researchers from Carnegie Mellon University published a document revealing the fact that many people did not know how to properly sign & encrypt messages by using the latest version of PGP encryption method. (SOURCE: www.lifehacker.com)

After a period of eight years, yet another group of researchers from the same university published a research paper to follow up the initial research, and revealed that despite the new version of PGP being easier to operate, the majority of individuals still had problems in signing and encrypting their messages. It also stated that finding & verifying public encryption keys of other people as well as sharing their own key was problematic and difficult for them.

Because of this, compliance and security managers at organizations as well as governmental agencies turn towards automated ways of encryption. Rather than depending on voluntary cooperation from employees, they develop automated encryption processes based on certain policies that take the decision and operate without user input. Once this is in place, emails are channeled through a gateway that has been configured to make sure it is complaint with the organization's security & regulatory policies. Those emails needing to be encrypted are automatically done so and transmitted.

Also, if the email is sent to a recipient who happens to work within the same organization, the email is automatically decrypted, making the process appear completely transparent to the users. However, those users who are not behind the luxury of such a gateway would need to take appropriate steps to ensure that their emails and properly encrypted or decrypted as needed.

# Chapter 11: Email Etiquette

Almost every organization that operates its own mail server should to have a clear email policy to educate the employees about what it considered to be an appropriate usage of email. It is extremely common for individuals to send content through email content to other people, and because some of these emails may contain inappropriate content, the name of the organization may be maligned, at the very least.

In most cases, a distinct email usage policy is developed and included in the organization's employee handbook for easy reference. It is crucial for organizations, system administrators, and other managers to make sure that this policy is implemented and practiced throughout the organization.

Some of the reasons why organizations need an email policy or etiquettes guidelines are listed below:

- **Personal email usage** – Employees must always know whether or not the emails they have created are acceptable. An excellent idea is to set a limit and assign a certain time of the day when emails of a personal nature can be sent. Additionally, it must be ensured that all the employees prevented from sending executable email attachments as these can become carriers for spreading of malware. Imposing a file size limit tends to be a good idea for personal usage of emails as they can clog up precious bandwidth.

- **Prohibited content** – It needs to be clearly communicated to the entire organization which content is prohibited. This may include – not limiting other factors, abusive content and offensive comments related to race, gender, religion, disability, or sexual orientation. Any individual who breaks these rules should be subjected to disciplinary action, leading to termination for serious offenses. Some organizations also prohibit use of email to discuss direct competitors or potential mergers or acquisitions.

- **Email risks** – A lot of people are not aware of the risks that email messages can carry. Therefore, it is crucial to develop etiquettes guidelines to educate them of all the possibilities.

- **Wastage of time** – Using a company's email system for non-productive activities imposes an unnecessary load on the server and hogs network traffic. Therefore, a usage policy may be needed to define and state that use of email should be strictly for office purposes.

- **Best practices** – Email etiquette has to be followed by each and every individual operating on a corporate server. This is crucial for safeguarding and maintaining a good reputation of an organization.

- **Reduced liability** – If some kind of incident occurs involving the employee, such as harassment or any other type of issue, an email policy can greatly reduce the liability of the company. There have been cases where an email policy has helped a company whose employee or employees were involved in some form of misconduct or illegal activity. If your email policy compels your users to avoid such activities, and they have signed it – then your organization will be completely free of any liability if any legal issues arise. There was a case of a company that faced legal action from two of its former employees for allowing racially offensive jokes to be circulated through the company's servers. However, the company defended itself by proving that they had a policy in place to deal with such cases and it prevented users from racism, and that strict and timely action had already been taken against the employees who were involved.

- **Email monitoring** – If you plan to monitor the emails that are sent and received, you will need to specifically mention this to your employees, otherwise the action would be considered a breach of privacy.

# Enforcing Your Email Policy

After you have developed an email policy, it is important to enforce it as well. Here are a few ways to help you do so.

## Provide Training

It is crucial to train the employees in your organization on how to adhere to the company's email policy. Assist users in learning to send effective mails by educating them about the best practices, while also explain that any offensive jokes or remarks can yield more harmful results than they think. It is very important to stress that the employees who witness any abuse of the company's email system should report it to their higher authority as soon as possible.

The employees should be taught not to open clicks or attachments from unsolicited emails nor provide their personal details to anyone whatsoever; regardless of how genuine it may look. If your email systems do not employ automatic encryption techniques, then they should be taught how to encrypt and decrypt emails successfully. Do not neglect to mention the use of digital signatures.

## Take Timely Action

If any employee lodges a complaint about any offensive or potentially harmful emails, it is absolutely imperative that the situation be dealt with in a timely manner. There should be internal procedures in place that allow an investigation into such incidents/complaints. Employees should also be encouraged to come forth if they feel as if they have been harassed in anyway. Timely action can save a company from huge legal costs, just as it was mentioned before. In the previous example, action was taken within 7 days of receiving the complaints from the two employees about the offensive content in emails. This led the supervisors to arrange two meetings to talk about the issue in detail.

The sender of the email was reprimanded and a written warning was placed in her employee file along with a verbal reproach. The company also asked its employees to re-review the company's policy and comply with it. All of this resulted in the court deeming that the employer had indeed 'acted reasonably' and the case against the company was dismissed.

## Monitor Emails

Emails should be monitored to ensure compliance with the policy is not being breached. The emails can be easily monitored when they are stored on the email server of the company. However, the best way is to place filters that automatically filter out and block emails containing unauthorized words/attachments. This would allow you to detect patterns of misuse and breach of policy. Nonetheless, make sure you include a privacy statement in your email policy as to how the emails will be monitored to prevent legal issues.

## Business Email Etiquettes

With more than 300 billion emails sent out each day, email etiquettes must be adhered to at work in order to ensure that this technology is used efficiently and appropriately in the business environment. Unfortunately, many people at workplaces often abuse this technology by using it in an inappropriate manner. This can not only hog your mail server bandwidth, but can also lead to an overall decrease in employee productivity.

Here are a few guidelines that you should pass on to the employees in your company to get them back on track when it comes to using the company's email system.

## 1. Send Clear, Concise Emails Only

The very first rule related to email usage states that long email should be avoided. There is no point of including irrelevant information in the first place. With that said, emails should be written in a clear, concise format where the topic stays onto the subject being discussed. Users should be instructed that if they are requesting for more information, they should clearly ask for it. If they want a certain decision to be made, they should ask for it. There is no point in getting people to wonder what the email is for in the first place.

## 2. Use a Meaningful Subject

The subject line of the email should never be left blank. It is absolutely important that a meaningful subject is included so that the content of the email is reflected in the subject line. Not only will this streamline the overall work process, but it will also remove the possibility of any confusion and will make it easy to locate older emails if the need arises.

## 3. Avoid Lavishly Formatted Emails

The emails should be formatted in a professional manner. Use of multicolored fonts and images should be avoided. Unnecessary use of images and pictures will hog the bandwidth, slowing down the email system, in some cases. The company may make it compulsory for the employees to use a particular look and feel; this practice should be adhered to.

Also, if the email is used to communicate with individuals outside of the company – such as clients – a company logo and a signature would enhance the professional appearance.

## 4. Respond in a Timely Manner

Users should be encouraged to reply to emails in a timely manner. Making people wait without a reason is unnecessary and portrays an unprofessional image of the company. For instance, if customers are made to wait for no good reason, the reputation of the company would be seriously affected. This is more of a courtesy rather than a rule – however, a company can implement it in its email usage policy if it deems it suitable and necessary.

# Finally

Setting up and running a mail server requires careful consideration, planning, as well as implementation. If an organization is planning to run its own email server, numerous things will have to be considered, including operational and legal aspects.

If you are entrusted with the task of designing your company's email architecture, then you will have to take into account the needs of your company, including the number of employees, security aspects, and reliability. Due to the fact that email servers have to process large amounts of data, reliability plays an immensely important role. An email server with poor hardware and incorrect software configuration will only make matters worse for the company, not to mention making it easier for cybercriminals to crack into the organization's personal data.

If, for instance, your organization wants to avoid the hassle of maintaining its own mail servers, then there are options out there that can do it for you. For instance, ZohoMail enables businesses to use their own domain, while it provides the email services including both storage and backup facilities.

Nonetheless, whatever approach you wish to take for your organization's email needs, do not forget to educate your employees on how they should be using this technology in a responsible and a safe way.

The purpose of this book was to provide an overview of the things involved when it comes to designing an email architecture and it implementation. It is hoped that you will use the knowledge gained to your advantage and will contribute to making this technology safer for everyone.

# Glossary of Email Terms

Source: http://www.lsoft.com/resources/glossary.asp

**Above the fold:** The top part of an email message that is visible to the recipient without the need for scrolling. The term originally comes from print and refers to the top half of a folded newspaper.

**Alias:** A unique and usually shorter URL (link) that can be distinguished from other links even if they ultimately go to the same Web page. This makes it possible to track which message led viewers to click on the link.

**ASP:** Application Service Provider – A company that offers organizations access over the Internet to applications and related services that would otherwise have to be located on site at the organization's premises.

**Attachment:** An audio, video or other data file that is attached to an email message.

**Autoresponder:** A computer program that automatically responds with a prewritten message to anyone who sends an email message to a particular email address or uses an online feedback form.

**Authentication:** A term that refers to standards, such as Sender ID, SPF and DomainKeys/DKIM, that serve to identify that an email is really sent from the domain name and individual listed as the sender. Authentication standards are used to fight spam and spoofing.

**B2B:** Business-to-Business – The exchange of information, products or services between two businesses – as opposed to between a business and a consumer (B2C).

**B2C:** Business-to-Consumer – The exchange of information, products or services between a business and a consumer – as opposed to between two businesses (B2B).

**Back-end:** (1) The part of the computer that changes source code into object code (machine read code). (2) The part of the program that runs on a server in a client/server application.

**Bayesian filter:** A spam filter that evaluates email message content to determine the probability that it is spam. Bayesian filters are adaptable and can learn to identify new patterns of spam by analyzing incoming email.

**BITNET:** Abbreviation for "Because It's Time NETwork." BITNET is primarily a network of sites for educational purpose and is separate from the Internet. Email is exchanged at no charge between BITNET and the Internet.

**Blacklist:** A list containing email addresses or IP addresses of suspected spammers. Blacklists are sometimes used to reject incoming mail at the server level before the email reaches the recipient.

**Block:** An action by an Internet Service Provider to prevent email messages from being forwarded to the end recipient.

**Bounces:** Email messages that fail to reach their intended destination. "Hard" bounces are caused by invalid email addresses, whereas "soft" bounces are due to temporary conditions, such as overloaded inboxes.

**CGI:** Common Gateway Interface – A specification for transferring information between a Web server and a CGI program. CGI programs are often used for processing email subscriptions and Web forms.

**Challenge-Response:** An authentication method that requires a human to respond to an email challenge message before the original email that triggered the challenge is delivered to the recipient. This method is sometimes used to cut down on spam since it requires an action by a human sender.

**Click-through tracking:** The process of tracking how many recipients clicked on a particular link in an email message. This is commonly done to measure the success of email marketing campaigns.

**Click-through rate:** In an email marketing campaign, the percentage of recipients who clicked on a particular link within the email message.

**Conditional blocks:** A text fragment that is pasted into an email message only if certain conditions are met (for instance the recipient lives in a certain area). Conditional blocks allow email marketers to create more personalized mailings.

**Conversion rate:** A measure of success for an email marketing campaign (for instance the number of recipients who completed a purchase). With email marketing, conversion rates are relatively easy to calculate because of the technology's measurable nature.

**CPM:** Cost Per Thousand – An industry standard measure for ad impressions. Email has a relatively low CPM compared to other marketing channels (Note: "M" represents thousand in Roman numerology).

**CRM:** Customer Relationship Management – The methodologies, software, and Internet capabilities that help a company manage customer relationships in an efficient and organized manner.

**Database Management System:** A database system that provides possibilities for users to connect to a database back-end and, hence, send out personalized messages to customers, according to their demographic information and preferences.

**Deliverability:** A term that refers to the best practices and authentication techniques of mass email communication that improve the likelihood that opt-in email messages are successfully delivered to end recipients instead of being erroneously blocked by ISPs and spam filters.

**Discussion group:** An email list community where members can obtain and share information. Every member can write to the list, and in doing so, everyone subscribed to the list will receive a copy of the message.

**DNS:** Domain Name Server (or system) – An Internet service that translates domain names into IP addresses.

**DomainKeys/DKIM:** DomainKeys/DomainKeys Identified Mail are cryptographic authentication solutions that add signatures to email messages, allowing recipient sites to verify that the message was sent by an authorized sender and was not altered in transit.

**Domain name:** A name that identifies one or more IP addresses. Domain names always have at least two parts that are separated by dots (for instance lsoft.com). The part on the left is the second-level domain (more specific), while the part on the right is the top-level domain (more general).

**Domain Throttling:** A technique that allows you to limit the number of email messages sent to a domain within a certain time frame. It is used to comply with ISPs and to avoid tripping spam filters. Many ISPs have their own policies and preferred limits.

**Double opt-in:** The recommended procedure for subscribing email recipients to an email list or newsletter. Once a person requests to subscribe to a list, a confirmation email message is automatically sent to the supplied email address asking the person to verify that they have in fact requested to be included in future mailings.

**Download:** To transfer a copy of a file from an Internet server to one's own computer.

**Email:** Email allows you to send and receive text, HTML, images and other data files over the Internet. Email is one of the most popular online activities and has become a vital tool for electronic commerce.

**Email bounces:** Email messages that fail to reach their intended destination. "Hard" bounces are caused by invalid email addresses, whereas "soft" bounces are due to temporary conditions, such as overloaded inboxes.

**Email client:** The software that recipients use to read email. Some email clients have better support for HTML email than others.

**Email harvesting:** The disreputable and often illegal practice of using an automated program to scan Web pages and collect email addresses for use by spammers.

**Email header:** The section of an email message that contains the sender's and recipient's email addresses as well as the routing information.

**Email marketing:** The use of email (or email lists) to plan and deliver permission-based marketing campaigns.

**False positive:** A legitimate email message that is mistakenly rejected or filtered by a spam filter.

**Forward DNS Lookup:** A Forward DNS Lookup, or just DNS Lookup, is the process of looking up and translating a domain name into its corresponding IP address. This can be compared to a Reverse DNS Lookup, which is the process of looking up and translating an IP address into a domain name.

**FQDN:** Fully Qualified Domain Name – A name consisting of both a host and a domain name. For example, www.lsoft.com is a fully qualified domain name. www is the host; lsoft is the second-level domain; and .com is the top-level domain.

**Freeware:** A free computer program usually made available on the Internet or through user groups.

**FTP:** File Transfer Protocol – Used for uploading or downloading files to and from remote computer systems on a network using TCP/IP, such as the Internet.

**Gateway:** This is a hardware or software set-up that functions as a translator between two dissimilar protocols. A gateway can also be the term to describe any mechanism providing access to another system (e.g AOL might be called a gateway to the Internet).

**Hard bounces:** Email messages that cannot be delivered to the recipient because of a permanent error, such as an invalid or non-existing email address.

**Host:** When a server acts as a host it means that other computers on the network do not have to download the software that this server carries.

**Host name:** The name of a computer on the Internet (such as www.lsoft.com).

**HTML:** HyperText Markup Language – The most commonly used coding language for creating Web pages. HTML can also be used in email messages.

**IMAP:** Internet Message Access Protocol – A protocol used to retrieve email messages. Most email clients use either the IMAP or the POP protocol.

**In-house list:** A list of email addresses that a company has gathered through previous customer contacts, Web sign-ups or other permission-based methods. In-house lists typically generate higher conversion rates than rented lists.

**Internet:** The largest worldwide computer network.

**Intranet:** Contrary to the public Internet, an intranet is a private network inside a company or organization.

**IP address:** An IP (Internet Protocol) address is a unique identifier for a computer on the Internet. It is written as four numbers separated by periods. Each number can range from 0 to 255. Before connecting to a computer over the Internet, a Domain Name Server translates the domain name into its corresponding IP address.

**ISP:** Internet Service Provider – A company that provides access to the Internet, including the World Wide Web and email, typically for a monthly fee.

**LAN:** Local Area Network, which is a computer network, although geographically limited, usually to the same building, office, etc.

**List broker:** A company that sells or rents lists of email addresses. Some list brokers are not reputable and sell lists with unusable or unsubstantiated candidates. It is therefore advisable for email marketers to build their own internal lists.

**List owner:** The owner of an email list defines the list's charter and policy (i.e. what the list is about and what are the general rules that all subscribers must accept in order to be subscribed to the list). The list owner is also responsible for administrative matters and for answering questions from the list subscribers.

**Mail-merge:** A process that enables the delivery of personalized messages to large numbers of recipients. This is usually achieved using email list management software working in conjunction with a database.

**Mainframe:** A high-level computer often shared by multiple users connected by individual terminals.

**Merge-purge:** The act of removing duplicate email addresses from a coalesced list that is composed of two or more existing lists.

**MIME:** Multi-Purpose Internet Mail Extensions – An extension of the original Internet email standard that allows users to exchange text, audio or visual files.

**Moderated list:** Moderators must approve any message posted to an email list before it is delivered to all subscribers. It is also possible for the moderator to edit or delete messages. A moderated list thus puts the list owner in the equivalent position as an editor of a newspaper.

**Multi-threading:** A process though which a mail server can perform multiple concurrent deliveries to different domains, which greatly speeds up the delivery of large volumes of email.

**Multipart/alternative:** A message format that includes both text and HTML versions. Recipients can then open the message in their preferred format.

**ODBC:** Open DataBase Connectivity – A Microsoft standard for accessing different database systems from Windows, for instance Oracle or SQL.

**Offload:** To assume part of the processing demand from another device.

**Open-relay:** Open-relay is the third-party relaying of email messages though a mail server. Spammers looking to obscure or hide the source of large volume mailings often use mail servers with open-relay vulnerabilities to deliver their email messages.

**Open-up tracking:** The process of tracking how many recipients opened their email messages as part of an email marketing campaign. Open-up tracking is only possible using HTML mail.

**Open-up rate:** The percentage of recipients who opened their email messages. The open-up rate is often used to measure the success of an email marketing campaign.

**Operating system:** A program that manages all other programs in a computer, such as Windows or Unix.

**Opt-in:** An approach to email lists in which subscribers must explicitly request to be included in an email campaign or newsletter.

**Opt-out:** An approach to email lists in which subscribers are included in email campaigns or newsletters until they specifically request not to be subscribed any longer. This method is not recommended and may in some cases be illegal.

**Out-of-office replies:** Automatic email reply messages triggered by incoming email to a user's inbox, typically activated when users are on vacation or otherwise unavailable through email for an extended period.

**Outsourcing:** An arrangement where one company provides services to another company that would otherwise have been implemented in-house (See also "ASP").

**Pass-along:** An email message that gets forwarded by a subscriber to another person who is not subscribed to the list (See also "Viral Marketing").

**Personalization:** The insertion of personal greetings in email messages (for instance "Dear John" rather than the generic

"Dear Customer"). Personalization requires email list management software that allows for so called mail-merge operations.

**Plain text:** Text in an email message that contains no formatting elements.

**POP:** Post Office Protocol – A protocol used to retrieve email from a mail server. Most email clients use either the POP or the newer IMAP protocol.

**Privacy:** A major concern of Internet users that largely involves the sharing of personally identifiable information, which includes name, birth date, Social Security number and financial data, for example.

**Protocol:** The set of formal rules that describe how to transmit data, especially across a network of computers.

**Query:** A subset of records in a database. Queries may be used to create highly specified demographics in order to maximize the effectiveness of an email marketing campaign.

**Reverse DNS Lookup:** A Reverse DNS Lookup is the process of looking up and translating an IP address into a domain name. This can be compared to a Forward DNS Lookup, which is the process of looking up and translating a domain name into its corresponding IP address.

**Rich media:** An Internet advertising term for a Web page that uses graphical technologies such as streaming video, audio files or other similar technology to create an interactive atmosphere with viewers.

**Router (Routing System):** The role of a router can be described as a bridge between two or more networks. The function of the router is to look at the destination addresses of the packets passing through it, and thereafter decide which route to send these packets on.

**Scalability:** The ability of a software program to continue to function smoothly as additional volume, or work is required of it.

**Sender ID:** Sender ID is an authentication protocol used to verify that the originating IP address is authorized to send email for the domain name declared in the visible "From" or "Sender" lines of the email message. Sender ID is used to prevent spoofing and to identify messages with visible domain names that have been forged.

**Server:** A program that acts as a central information source and provides services to programs in the same or other computers. The term can either refer to a particular piece of software, such as a WWW server, or to the machine on which the software is running.

**Shareware:** This term refers to software available on public networks.

**Signature file:** A short text file that email users can automatically append at the end of each message they send. Commonly, signature files list the user's name, phone number, company, company URL, etc.

**SMTP:** Simple Mail Transfer Protocol – A protocol used to send email on the Internet. SMTP is a set of rules regarding the interaction between a program sending email and a program receiving email.

**Snail mail:** Traditional or surface mail sent through postal services such as the USPS.

**Soft bounces:** Email messages that cannot be delivered to the recipient because of a temporary error, such as a full mailbox.

**Spam:** (Also known as unsolicited commercial email) – Unwanted, unsolicited junk email sent to a large number of recipients.

**SPF:** Sender Policy Framework – An authentication protocol used by recipient sites to verify that the originating IP address is authorized to send email for the domain name declared in the "MAIL FROM" line of the mail envelope. SPF is used to identify messages with forged "MAIL FROM" addresses.

**Spoofing** The disreputable and often illegal act of falsifying the sender email address to make it appear as if an email message came from somewhere else.

**Streaming media:** Audio and video files transmitted on the Internet in a continuous fashion.

**Subject line:** The part of an email message where senders can type what the email message is about. Subject lines are considered important by email marketers because they can often influence whether a recipient will open an email message.

**Targeting:** Using demographics and related information in a customer database to select the most appropriate recipients for a specific email campaign.

**TCP / IP:** Transmission Control Protocol / Internet Protocol – This is the protocol that defines the Internet. TCP / IP was originally designed for the unix operating system, but is today available for every major kind of computer operating system.

**Tracking:** In an email marketing campaign, measuring behavioral activities such as click-throughs and open-ups.

**URL:** Uniform Resource Locator – The address of a file or Web page accessible on the Internet

**Viral marketing:** A marketing strategy that encourages email recipients to pass along messages to others in order to generate additional exposure.

**Virtual hosting:** A Web server hosting service that replaces a company's need to purchase and maintain its own Web server and connections to the Internet.

**Virus:** A program, macro or fragment of code that causes damage and can be quickly spread through Web sites or email.

**Whitelist:** A list of pre-authorized email addresses from which email messages can be delivered regardless of spam filters.

**Worm:** Malicious code that is often spread through an executable attachment in an email message.

**XML:** Extensible Markup Language – A flexible way to create standard information formats and share both the format and the data on the World Wide Web.

Made in the USA
San Bernardino, CA
11 October 2016